Advanced Cell and Molecular Biology

A Deep Dive into Cutting-Edge Research, Molecular Techniques, and Emerging Frontiers in Cell Science

Written by

Bolakale Aremu

Advanced Cell and Molecular Biology

A Deep Dive into Cutting-Edge Research, Molecular Techniques, and Emerging Frontiers in Cell Science

Published in the United States

ISBN: 979-8-3492-5100-9

Disclaimer

Table of Contents

0. Preface

0.1. About This Book

Unlock the Secrets of the Cell—Beyond the Basics! Are you ready to move beyond foundational biology and dive into the cutting edge of modern science? *Advanced Cell and Molecular Biology* is your gateway to the next level of understanding.

Crafted for upper-level students, researchers, and professionals, this advanced guide explores the intricate molecular mechanisms that govern life, the transformative technologies revolutionizing biological research, and the discoveries shaping the future of medicine, genetics, and biotechnology.

From CRISPR gene editing and chromatin remodeling to single-cell analysis, synthetic biology, and cancer cell signaling, each chapter provides a research-informed deep dive into the molecular heartbeat of cells.

What sets this book apart is its **modular and concise structure**, carefully designed to maximize clarity and learning efficiency without sacrificing depth. The bullet-point format and focused subtopics allow readers to quickly grasp complex ideas, making the book ideal for fast reference, exam prep, or on-the-go research support.

In this book you'll find:

- Clear, concise explanations of complex biological processes
- High-quality illustrations and diagrams for enhanced understanding
- Up-to-date research insights and real-world scientific applications
- A modular chapter layout for targeted, flexible learning
- Compact sections that support both deep study and quick review

Whether you're preparing for graduate school, conducting cutting-edge research, or working in the biotech or biomedical industries, this book will sharpen your expertise and expand your scientific perspective.

Now you can explore the frontier of life science, and master the complexity of the cell—with precision, clarity, and insight.

Key Features of This Book:

- **Advanced Focus**: Designed specifically for readers with prior knowledge of basic biology and molecular science.
- **Cutting-Edge Topics**: Covers recent developments in gene regulation, systems biology, therapeutic biotechnology, and more.
- **Techniques and Tools**: Introduces the latest experimental and analytical techniques in modern cell biology research.
- **Relevant Visuals**: Includes one or two high-quality images to support complex concepts and experiments.
- **Modular Structure**: Easy-to-follow numbering system for seamless navigation and referencing.
- **Real-World Applications**: Emphasizes how molecular biology is applied in medical, industrial, and academic contexts.

Whether you're advancing your education, conducting research, or simply passionate about biological science, this book offers a valuable roadmap through the exciting and ever-evolving world of cell and molecular biology.

0.2. Abbreviations and Terminology

To aid understanding throughout this book, below is a list of common abbreviations and specialized terms frequently used in the field of advanced cell and molecular biology. While many terms are defined in context, this section serves as a quick reference guide.

0.2.1. Common Abbreviations

Abbreviation	Full Term
DNA	Deoxyribonucleic Acid
RNA	Ribonucleic Acid
mRNA	Messenger RNA
tRNA	Transfer RNA
rRNA	Ribosomal RNA
miRNA	MicroRNA
siRNA	Small Interfering RNA
ATP	Adenosine Triphosphate
ADP	Adenosine Diphosphate

Abbreviation	Full Term
NAD^+	Nicotinamide Adenine Dinucleotide

NADH	Reduced Nicotinamide Adenine Dinucleotide
FAD	Flavin Adenine Dinucleotide
cAMP	Cyclic Adenosine Monophosphate
PCR	Polymerase Chain Reaction
qPCR	Quantitative PCR
CRISPR	Clustered Regularly Interspaced Short Palindromic Repeats
Cas9	CRISPR-associated protein 9
ER	Endoplasmic Reticulum

Abbreviation	Full Term
RER	Rough Endoplasmic Reticulum
ECM	Extracellular Matrix
GFP	Green Fluorescent Protein
SDS-PAGE	Sodium Dodecyl Sulfate Polyacrylamide Gel Electrophoresis
ELISA	Enzyme-Linked Immunosorbent Assay
FISH	Fluorescence In Situ Hybridization
TEM	Transmission Electron Microscopy
SEM	Scanning Electron Microscopy

NGS	Next-Generation Sequencing
RT	Reverse Transcription
RT-PCR	Reverse Transcription Polymerase Chain Reaction
siRNA	Small Interfering RNA
snRNA	Small Nuclear RNA
UTR	Untranslated Region
ORF	Open Reading Frame

Key Terminology

- **Apoptosis**: Programmed cell death, a regulated process to eliminate damaged or unnecessary cells.
- **Epigenetics**: The study of heritable changes in gene function that do not involve changes in the DNA sequence.
- **Transcription**: The process by which a segment of DNA is copied into RNA.
- **Translation**: The synthesis of proteins based on the sequence encoded in mRNA.

- **Gene Expression**: The process through which genetic information is used to produce proteins or RNA molecules.
- **Signal Transduction**: The process by which a cell responds to signals from its environment through a series of molecular events.
- **Genome Editing**: Techniques such as CRISPR-Cas9 that enable precise alterations in the DNA of living organisms.
- **Synthetic Biology**: The design and construction of new biological parts, devices, and systems.
- **Molecular Cloning**: Techniques used to replicate and manipulate DNA sequences in laboratory settings.
- **Proteomics**: The large-scale study of proteins, particularly their structures and functions.
- **Transcriptomics**: The study of RNA transcripts produced by the genome under specific circumstances.

Part I: Foundations of Advanced Cell Biology

1. Cellular Architecture and Function Revisited

1.1. Prokaryotic vs. Eukaryotic Cells

Understanding the fundamental differences between prokaryotic and eukaryotic cells is essential before diving into advanced molecular and cellular mechanisms. While these two cell types share many basic features, they differ significantly in structure, complexity, and regulatory mechanisms—differences that are critical in both research and applied biosciences.

1.1.1. Overview of Cell Types

Prokaryotic Cells
These are simple, unicellular organisms that lack a nucleus and membrane-bound organelles. Their genetic material is contained in a single circular DNA molecule located in the nucleoid region. Bacteria and Archaea are the primary domains of life that include prokaryotic organisms.

Eukaryotic Cells
Eukaryotic cells are structurally complex and may be unicellular or multicellular. They possess a true nucleus enclosed in a nuclear envelope, as well as numerous membrane-bound organelles such as mitochondria, endoplasmic reticulum, and Golgi apparatus. Eukaryotes include protists, fungi, plants, and animals.

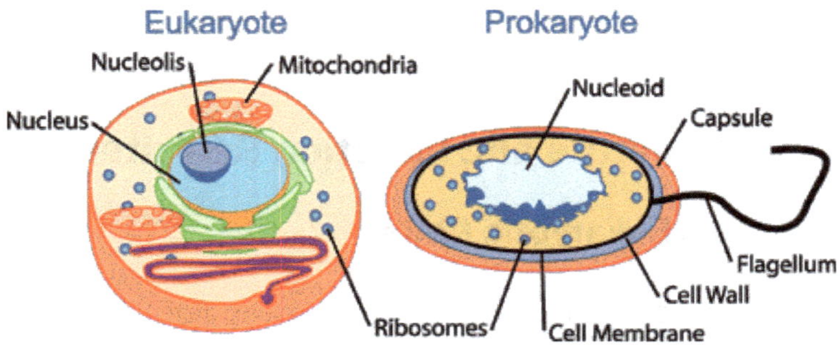

Figure 1.1: Diagram showing key structural differences between prokaryotic and eukaryotic cells, highlighting nucleus, organelles, and cell wall. Source: Wikimedia. Link: https://commons.wikimedia.org/wiki/File:Celltypes.png

1.1.2. Comparative Table: Prokaryotes vs. Eukaryotes

Feature	**Prokaryotic Cells**	**Eukaryotic Cells**
Nucleus	Absent	Present
DNA Form	Circular (naked)	Linear, associated with histones
Organelles	None (except ribosomes)	Membrane-bound organelles present
Cell Division	Binary fission	Mitosis or meiosis
Size	Typically 0.1–5 μm	Typically 10–100 μm

Ribosomes	70S	80S (cytoplasmic), 70S (mitochondria/chloroplasts)
Cell Wall	Peptidoglycan (in bacteria)	Cellulose (plants), chitin (fungi), absent in animals
Examples	*E. coli, Streptococcus, Archaea*	*Yeast, Paramecium, human cells*

1.1.3. Structural and Functional Implications

- **Genome Organization**: Prokaryotes typically contain a single, compact genome, whereas eukaryotes have multiple chromosomes within a nucleus. This impacts gene regulation, replication, and epigenetics.
- **Compartmentalization**: Eukaryotic organelles allow for specialized environments, increasing the efficiency and regulation of cellular processes such as energy production and protein synthesis.
- **Cellular Communication**: Eukaryotic cells use complex signaling networks and structures like vesicles and cytoskeletal tracks, while prokaryotes rely on simpler but highly responsive signal transduction pathways.
- **Gene Regulation**: Prokaryotic gene expression is often polycistronic (multiple genes per transcript), whereas eukaryotic expression is usually monocistronic and heavily regulated via enhancers, silencers, and chromatin remodeling.

1.1.4. Scientific Insight: Evolutionary Perspective

The endosymbiotic theory provides a compelling explanation for the origin of mitochondria and chloroplasts in eukaryotic cells, suggesting that these organelles were once free-living prokaryotes. This theory is supported by the presence of double membranes, 70S ribosomes, and circular DNA within these organelles.

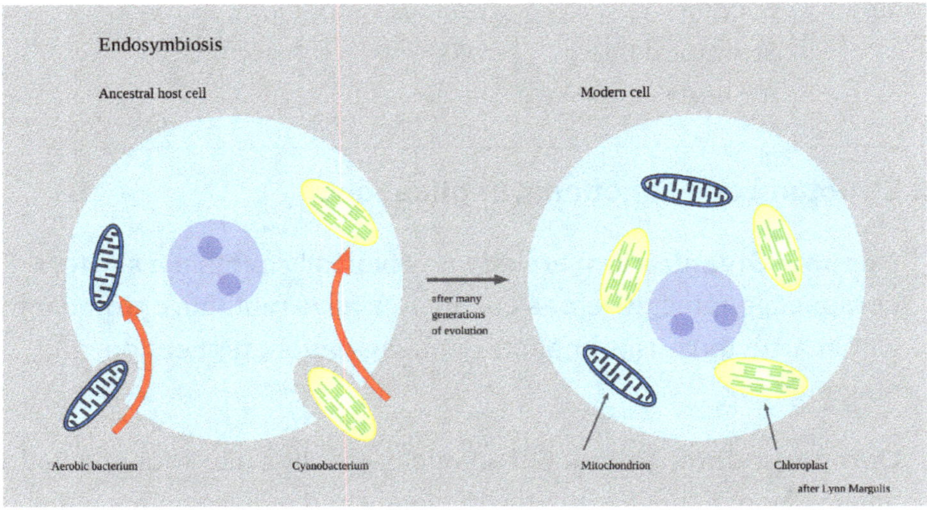

Figure 1.2: Visual representation of the endosymbiotic theory, depicting engulfment of a prokaryote by an ancestral eukaryotic cell. Source: Wikimedia. Link: https://commons.wikimedia.org/wiki/File:Endosymbiosis.svg

1.2. Organelles and Subcellular Compartments

The complexity and specialization of eukaryotic cells are largely due to their internal organization into membrane-bound organelles and subcellular compartments. These structures allow spatial and functional compartmentalization, enabling a multitude of biochemical reactions to occur simultaneously and efficiently. Understanding the function, structure, and dynamic interactions of these compartments is fundamental to advanced cell biology.

1.2.1. Major Organelles and Their Functions

Figure 1.3: A labeled diagram of a typical eukaryotic cell showing key organelles and their locations. Source: Wikimedia. Link: https://commons.wikimedia.org/wiki/File:Animal_Cell.svg

Organelle	Structure	Function
Nucleus	Double membrane with nuclear pores	Stores genetic material, site of DNA replication and transcription
Mitochondria	Double membrane with inner folds (cristae)	ATP production via oxidative phosphorylation, apoptosis regulation
Endoplasmic Reticulum (ER)	Network of membranous tubules	Rough ER synthesizes proteins; Smooth ER synthesizes lipids, detoxifies drugs

Golgi Apparatus	Flattened membranous sacs (cisternae)	Modifies, sorts, and packages proteins and lipids for secretion or delivery
Lysosomes	Membrane-enclosed vesicles with hydrolytic enzymes	Intracellular digestion of macromolecules and organelles
Peroxisomes	Small, membrane-bound vesicles	Breakdown of fatty acids and detoxification of hydrogen peroxide
Chloroplasts (in plants)	Double membrane, thylakoid stacks (grana)	Photosynthesis—conversion of light energy into chemical energy
Vacuoles	Large vesicles (prominent in plants)	Storage, waste disposal, turgor pressure maintenance
Cytoskeleton	Network of microtubules, microfilaments, and intermediate filaments	Maintains cell shape, enables movement and intracellular transport
Ribosomes	Composed of rRNA and proteins; free or ER-bound	Protein synthesis (translation)

1.2.2. Subcellular Compartments and Functional Zonation

- **Nucleolus**: Dense region within the nucleus where ribosomal RNA (rRNA) is synthesized and ribosomal subunits are assembled.
- **Endosomes**: Involved in the sorting and trafficking of internalized materials from the plasma membrane.
- **Transport Vesicles**: Shuttle proteins and lipids between organelles, often coated with proteins like clathrin or COPI/COPII.
- **Cytoplasm vs. Cytosol**: The cytoplasm includes all cell contents between the plasma membrane and the nuclear envelope, while the cytosol is the fluid portion excluding organelles.

1.2.3. Dynamic Interactions Between Organelles

Organelles are not isolated. They interact through:

- **Vesicular Transport**: Proteins and lipids are transferred between organelles via membrane-bound vesicles (e.g., from ER to Golgi).
- **Membrane Contact Sites (MCSs)**: Direct physical contacts between organelles (e.g., ER–mitochondria) facilitate lipid transfer and signaling.
- **Organelle Biogenesis and Turnover**: Organelles are dynamically assembled, disassembled, and recycled based on cellular needs (e.g., mitophagy for mitochondria).

1.2.4. Organelle Dysfunction and Disease

Defects in organelle structure or function can lead to severe diseases:

- **Lysosomal Storage Diseases**: Caused by enzyme deficiencies leading to accumulation of undigested molecules (e.g., Tay-Sachs, Gaucher disease).
- **Mitochondrial Disorders**: Affect energy metabolism and can manifest as muscle weakness, neurodegeneration, or cardiomyopathy.
- **ER Stress**: Leads to unfolded protein response (UPR), linked to diseases such as diabetes, cancer, and neurodegeneration.

1.3. Cytoskeleton: Structure and Dynamics

The cytoskeleton is a complex and dynamic network of protein filaments that extends throughout the cytoplasm of eukaryotic cells. Far from being a static scaffold, it provides structural support, facilitates intracellular transport, and enables cell motility, shape changes, and division. Understanding the cytoskeleton's architecture and behavior is fundamental to advanced cell biology and biomedical research.

Types of Cytoskeletal Filaments

The cytoskeleton is composed of three major classes of protein filaments, each with distinct structures, functions, and associated proteins:

Filament Type	Protein Subunits	Diameter	Primary Functions
Microfilaments	Actin	~7 nm	Cell shape, motility, cytokinesis, endocytosis

Intermediate filaments	Various (e.g., keratin, vimentin)	~10 nm	Mechanical strength, organelle positioning
Microtubules	Tubulin (α- and β-tubulin)	~25 nm	Intracellular transport, mitosis, cilia and flagella

1.3.1. Microfilaments (Actin Filaments)

Actin filaments are dynamic structures composed of actin monomers (G-actin) that polymerize to form filamentous actin (F-actin). They are highly concentrated beneath the plasma membrane and form structures such as lamellipodia, filopodia, and stress fibers.

Functions:

- Provide mechanical support to the cell cortex
- Drive cell motility and migration
- Participate in muscle contraction (in association with myosin)
- Involved in endocytosis and cell division

1.3.2. Intermediate Filaments

Intermediate filaments provide tensile strength and structural integrity. Unlike actin filaments and microtubules, they are more stable and less dynamic. Different cell types express different intermediate filament proteins (e.g., keratins in epithelial cells, neurofilaments in neurons).

Functions:

- Reinforce cell shape
- Anchor organelles in place
- Involved in nuclear architecture (nuclear lamins)

1.3.3. Microtubules

Microtubules are hollow tubes formed by the polymerization of α- and β-tubulin dimers. They originate from microtubule organizing centers (MTOCs) such as the centrosome.

Functions:

- Form mitotic spindle fibers during cell division
- Serve as tracks for motor proteins (kinesins and dyneins) for intracellular transport
- Support the structure of cilia and flagella

1.3.4. Cytoskeletal Dynamics and Regulation

Cytoskeletal filaments are highly dynamic, undergoing continuous polymerization and depolymerization. This dynamic behavior is tightly regulated by:

- Nucleation factors (e.g., Arp2/3 complex for actin, γ-tubulin for microtubules)
- Severing proteins (e.g., cofilin, katanin)
- Capping proteins that stabilize filament ends
- Motor proteins (e.g., myosin, kinesin, dynein) that convert ATP into mechanical work

These regulators allow rapid reorganization of the cytoskeleton in response to environmental cues, cell cycle progression, or signaling events.

1.3.5. Crosstalk Between Cytoskeletal Systems

The actin filaments, intermediate filaments, and microtubules are not independent systems. Instead, they are interconnected through linker proteins (e.g., plectin) and signaling pathways that coordinate their organization and functions. This crosstalk is essential for integrated cellular processes like migration, adhesion, and morphogenesis.

1.4. Cellular Membranes and Transport Mechanisms

Cellular membranes are essential structures that define boundaries, compartmentalize cellular functions, and regulate the transport of substances in and out of cells and organelles. Understanding their composition and the mechanisms of molecular transport is foundational to advanced studies in cell biology, physiology, and pharmacology.

1.4.1. Structure of Biological Membranes

Biological membranes primarily consist of a **phospholipid bilayer** interspersed with **proteins**, **cholesterol**, and **carbohydrates**. The amphipathic nature of phospholipids—having both hydrophilic (water-attracting) heads and hydrophobic (water-repelling) tails—leads to the spontaneous formation of bilayers in aqueous environments.

Key components:

- **Phospholipids**: Form the fundamental bilayer structure.
- **Integral (transmembrane) proteins**: Span the membrane and mediate transport, signal transduction, or adhesion.
- **Peripheral proteins**: Attach to the inner or outer surfaces and assist in signaling or structural support.

- **Cholesterol**: Modulates fluidity and stability.
- **Glycolipids and glycoproteins**: Involved in cell recognition and signaling.

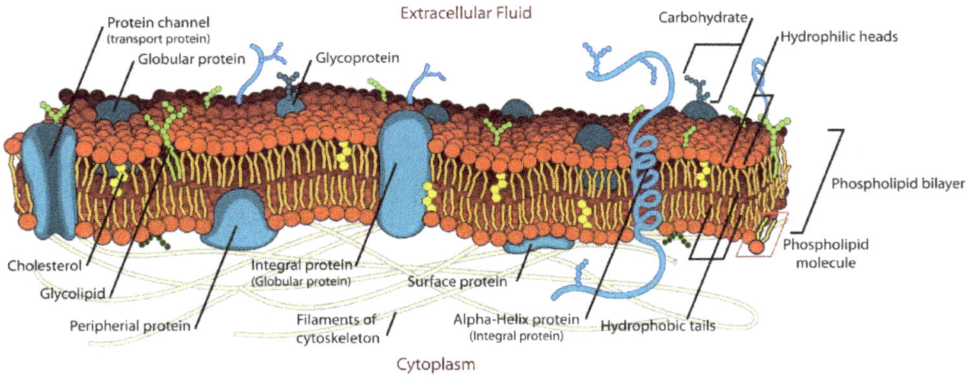

Figure 1.4: Diagram showing the phospholipid bilayer with embedded proteins, cholesterol, and surface carbohydrates. Source: Wikimedia. Link: https://commons.wikimedia.org/wiki/File:Cell_membrane_detailed_diagram_en.svg

The **fluid mosaic model**, first proposed by Singer and Nicolson in 1972, illustrates the dynamic and flexible nature of membranes, where proteins and lipids can move laterally within the bilayer.

1.4.2. Membrane Dynamics and Fluidity

Membrane fluidity is influenced by:

- **Fatty acid composition**: Unsaturated fatty acids increase fluidity.
- **Cholesterol content**: Acts as a buffer, maintaining fluidity at varying temperatures.
- **Temperature**: Higher temperatures increase membrane fluidity.

Cells tightly regulate fluidity to ensure proper function of membrane proteins and facilitate cellular processes like vesicle fusion, endocytosis, and signaling.

1.4.3. Membrane Transport Mechanisms

Cells exchange materials with their environment through various transport mechanisms, classified as **passive** or **active**.

A. Passive Transport

- **Simple diffusion**: Movement of small, nonpolar molecules (e.g., O_2, CO_2) across the membrane without assistance.
- **Facilitated diffusion**: Movement of polar or charged molecules via specific transport proteins (e.g., glucose via GLUT transporters).
- **Osmosis**: Diffusion of water across a selectively permeable membrane through aquaporins.

B. Active Transport

- Requires **ATP** or an electrochemical gradient to move substances **against** their concentration gradient.
- **Primary active transport**: Uses ATP directly (e.g., Na^+/K^+ ATPase).
- **Secondary active transport**: Couples the movement of one substance down its gradient to move another against its gradient (e.g., Na^+/glucose symporter).

1.4.4. Bulk Transport: Endocytosis and Exocytosis

Beyond individual molecules, cells move large particles and fluids via **vesicular transport** mechanisms:

Endocytosis: Uptake of materials by engulfing them in vesicles.

- **Phagocytosis**: "Cell eating" – ingestion of large particles or cells.
- **Pinocytosis**: "Cell drinking" – uptake of extracellular fluid and solutes.
- **Receptor-mediated endocytosis**: Selective uptake of specific molecules via receptor binding (e.g., LDL uptake).

Exocytosis: Fusion of vesicles with the plasma membrane to release contents outside the cell.

- Common in neurotransmitter release, hormone secretion, and membrane protein delivery.

1.4.5. Specialized Membrane Domains

Cells often create membrane subdomains for localized functions:

- **Lipid rafts**: Microdomains enriched in cholesterol and sphingolipids involved in signaling.
- **Tight junctions**: Prevent paracellular leakage in epithelial tissues.
- **Gap junctions**: Facilitate intercellular communication via small molecule exchange.

Understanding the sophisticated design and functionality of cellular membranes and transport systems is critical for exploring how cells maintain homeostasis, communicate, respond to stimuli, and interact with their environment—core principles in biotechnology, pharmacology, and disease pathogenesis.

2. Cell Communication and Signaling Pathways

2.1. Signal Transduction Mechanisms

Cell communication is a fundamental process by which cells perceive and respond to external and internal signals. This communication is crucial for coordinating development, immune responses, homeostasis, and tissue repair. The process of converting an extracellular signal into a functional change inside the cell is known as **signal transduction**.

2.1.1. Overview of Cell Signaling

Signal transduction involves a **ligand** (signaling molecule) binding to a **receptor**, triggering a cascade of molecular events inside the cell. These events amplify the signal, regulate gene expression, and alter cellular behavior.

Types of signaling based on distance:

- **Autocrine**: Cell signals to itself.
- **Juxtacrine (signaling across gap junctions)**: Direct contact between neighboring cells.
- **Paracrine**: Signal affects nearby cells.
- **Endocrine**: Hormones travel through the bloodstream to distant targets.

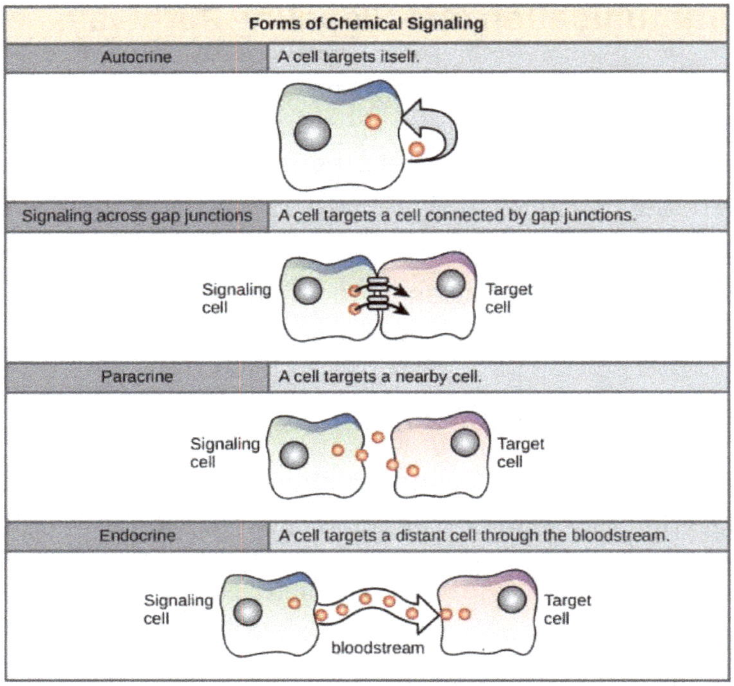

Figure 2.1: Illustration showing autocrine, paracrine, endocrine, and juxtacrine signaling mechanisms. Source: Openstax. Link: https://openstax.org/books/biology-2e/pages/9-1-signaling-molecules-and-cellular-receptors

2.1.2. Receptors and Ligands

Receptors are specialized proteins that recognize and bind ligands with high specificity.

Main classes of receptors:

- **G protein-coupled receptors (GPCRs)**: Activate G proteins to trigger downstream effects (e.g., adrenaline signaling).
- **Receptor tyrosine kinases (RTKs):** Phosphorylate themselves and other proteins upon ligand binding (e.g., insulin receptor).

- **Ion channel-linked receptors**: Open or close in response to ligands (e.g., neurotransmitter receptors).
- **Intracellular receptors**: Located in the cytoplasm or nucleus; respond to hydrophobic ligands like steroid hormones.

2.1.3. Signal Transduction Cascades

Signal transduction pathways often involve multiple intermediates and steps to amplify the response.

Key elements:

- **Second messengers**: Small molecules like cAMP, Ca^{2+}, IP_3 that relay signals inside the cell.
- **Protein kinases**: Enzymes that phosphorylate target proteins to activate or deactivate them.
- **Scaffold proteins**: Organize pathway components to ensure signal specificity.
- **Feedback mechanisms**: Regulate signal duration and intensity.

Example: cAMP Pathway (GPCR signaling)

1. Ligand binds to GPCR.
2. GPCR activates G protein.
3. G protein activates adenylyl cyclase → converts ATP to cAMP.
4. cAMP activates Protein Kinase A (PKA).
5. PKA phosphorylates downstream targets.

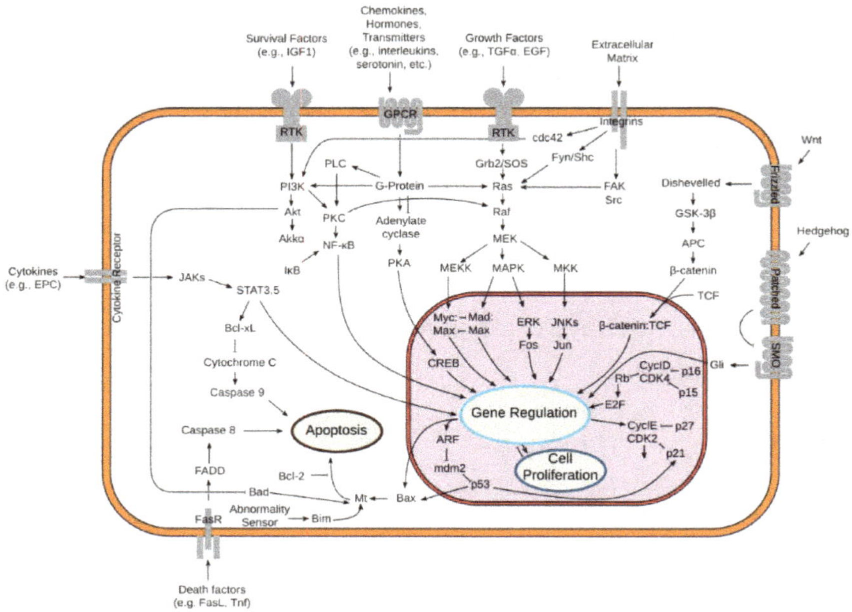

Figure 2.2: Diagram illustrating GPCR activation, G protein signaling, cAMP generation, and PKA activation. Source: Wikimedia. Link: https://commons.wikimedia.org/wiki/File:Signal_transduction_pathways.svg

2.1.4. Signal Amplification and Specificity

- **Amplification**: One receptor-ligand interaction can generate hundreds of secondary messengers, amplifying the response.
- **Specificity**: Achieved through spatial organization of signaling components, selective expression of receptors, and feedback loops.

2.1.5. Termination of Signaling

To avoid overstimulation or uncontrolled responses, cells deactivate signals through:

- **Receptor internalization or degradation**
- **Dephosphorylation of proteins by phosphatases**

- **Destruction of second messengers (e.g., cAMP hydrolyzed by phosphodiesterase)**

Signal transduction mechanisms form the foundation of many physiological processes and are central to understanding diseases like cancer, diabetes, and autoimmune disorders. Therapeutic targeting of specific signaling molecules has become a major focus in modern drug development.

2.2. Receptors and Secondary Messengers

Cellular communication relies heavily on the ability of cells to detect and respond to signals via specialized proteins called **receptors**, and to amplify those signals through **secondary messengers**. This dynamic system enables cells to maintain homeostasis, coordinate responses, and adapt to environmental changes.

2.2.1. Receptors: The Signal Sensors

Receptors are **proteins, typically located on the cell surface or within the cell**, that recognize and bind specific signaling molecules (ligands). Binding induces a conformational change that initiates downstream signaling pathways.

Receptors are generally classified based on their structure and location:

1. Cell Surface Receptors (Membrane-bound)

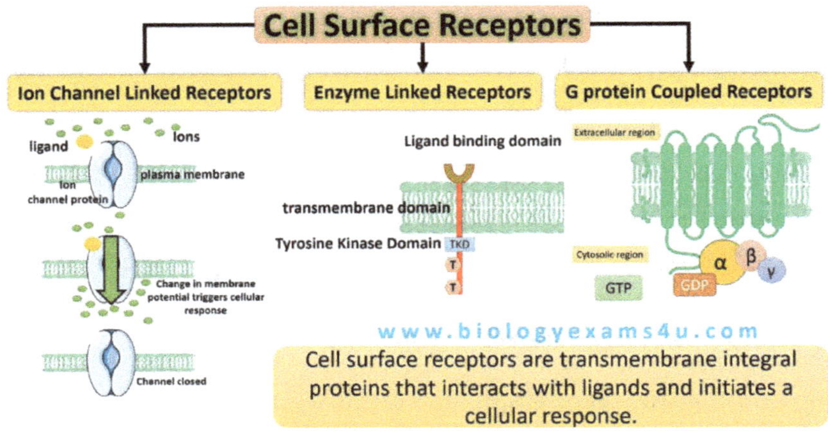

Figure 2.3: Three types of cell surface receptors. Source: BiologyExams4u. Link: https://www.biologyexams4u.com/2022/10/3-types-of-cell-surface-receptors.html

Detect **hydrophilic or large ligands** that cannot cross the lipid bilayer.

Examples:

- **G Protein-Coupled Receptors (GPCRs)**
- **Receptor Tyrosine Kinases (RTKs)**
- **Ion Channel-Linked Receptors**

2. Intracellular Receptors:

- Detect **small, lipophilic molecules** that can diffuse through membranes.
- Often act as **transcription factors** once activated.
- Example: **Steroid hormone receptors**

2.2.2. G Protein-Coupled Receptors (GPCRs)

- Among the most abundant and diverse receptors.
- Transduce signals via **heterotrimeric G proteins (α, β, γ subunits)**.

- Activation causes GDP-GTP exchange on the Gα subunit, triggering downstream effectors like **adenylyl cyclase**.

2.2.3. Receptor Tyrosine Kinases (RTKs)

- Ligand binding induces **dimerization and autophosphorylation**.
- Phosphorylated tyrosine residues recruit signaling proteins with SH2 domains.
- Critical in **growth factor signaling** (e.g., EGF, insulin receptors).

2.2.4. Secondary Messengers: Signal Amplifiers

Secondary messengers are **small molecules or ions** generated in response to receptor activation. They **propagate and amplify the signal** within the cell, often initiating multiple downstream pathways.

Common secondary messengers include:

1. Cyclic AMP (cAMP):

- Generated by **adenylyl cyclase** from ATP.
- Activates **protein kinase A (PKA)**, regulating metabolism and gene expression.

2. Cyclic GMP (cGMP):

- Activates **protein kinase G (PKG)**.
- Involved in **vasodilation** and phototransduction.

3. Inositol Triphosphate (IP_3) and Diacylglycerol (DAG):

- Produced via **phospholipase C** cleavage of PIP_2.
- IP_3 triggers **Ca^{2+} release** from the ER; DAG activates **protein kinase C**

(PKC).

4. Calcium Ions (Ca^{2+}):

- Released from ER or extracellular space.
- Bind to **calmodulin** and other effectors, influencing muscle contraction, secretion, and gene expression.

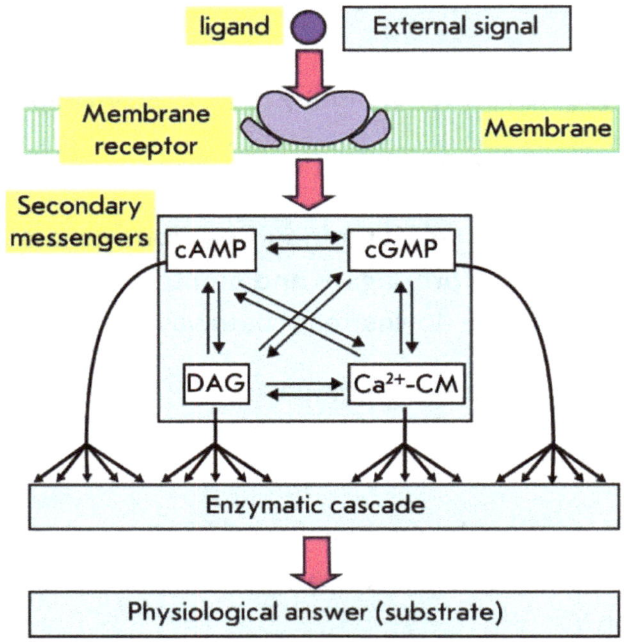

Figure 2.4: Flowchart showing how ligand binding to a receptor activates secondary messengers like cAMP, IP_3, DAG, and Ca^{2+}. Source: ResearchGate. Link: https://www.researchgate.net/figure/The-main-stages-of-signal-transduction-mediated-by-secondary-messengers-in-the-cell-cAMP_fig1_225079764

2.2.5. Spatial and Temporal Dynamics

1. Receptor and second messenger activity is **highly regulated**.

2. Signal termination is equally critical and involves:

- **Phosphodiesterases** breaking down cAMP/cGMP.
- **Phosphatases** removing phosphate groups.
- **Receptor internalization or desensitization**.

2.3. Crosstalk Between Pathways

Cellular signaling is rarely linear. Instead, cells integrate signals from multiple pathways through a phenomenon known as **signaling crosstalk**. This crosstalk ensures that signals are fine-tuned, coordinated, and context-dependent—allowing the cell to respond precisely to its environment.

2.3.1. What is Crosstalk?

Crosstalk refers to the interaction between two or more signaling pathways that modifies the output of each. These interactions can be **synergistic, antagonistic, or regulatory**, contributing to signal diversification and cellular decision-making.

Crosstalk allows cells to:

- Prioritize one pathway over another
- Integrate different extracellular cues
- Prevent overactivation or misactivation of signals

2.3.2. Mechanisms of Crosstalk

Crosstalk can occur at various levels within signaling pathways:

1. Shared Signaling Proteins:
Some proteins like **MAP kinases**, **PI3K**, or **Akt** are involved in multiple signaling routes, acting as integration points.

2. Competing Effectors:
Multiple pathways may compete for the same downstream targets, such as transcription factors (e.g., NF-κB, CREB).

3. Feedback Loops and Inhibition:

- **Positive feedback** enhances the signal (e.g., ERK activation upregulates RTK signaling).
- **Negative feedback** can inhibit other pathways to limit signal spread.

4. Co-receptors and Scaffold Proteins:
These organize components from different pathways into functional hubs, ensuring specific outcomes.

2.3.3. Examples of Functional Crosstalk

- **EGF and Insulin Signaling:**
  Epidermal Growth Factor Receptor (EGFR) signaling via RTK can interact with **PI3K-Akt** and **MAPK pathways**, as does insulin signaling—enabling coordinated responses in growth and metabolism.

- **GPCR and RTK Interactions:**
  GPCRs can transactivate RTKs, especially in the cardiovascular and immune systems, through **metalloproteinase-mediated cleavage** of RTK ligands.

- **Wnt and TGF-β Crosstalk in Development and Cancer:**
  These pathways share common transcriptional co-activators and repressors, leading to cooperative control of cell proliferation and differentiation.

2.3.4. Biological Significance of Crosstalk

- **Precision in Cellular Decisions:**
  Ensures the right gene expression patterns in response to complex environments.

- **Prevention of Signal Redundancy or Overactivation:**
  Crosstalk can suppress irrelevant or harmful pathways under specific conditions.

- **Adaptation and Learning in Cells:**
  In neurons and immune cells, crosstalk enables adaptive responses to stimuli over time.

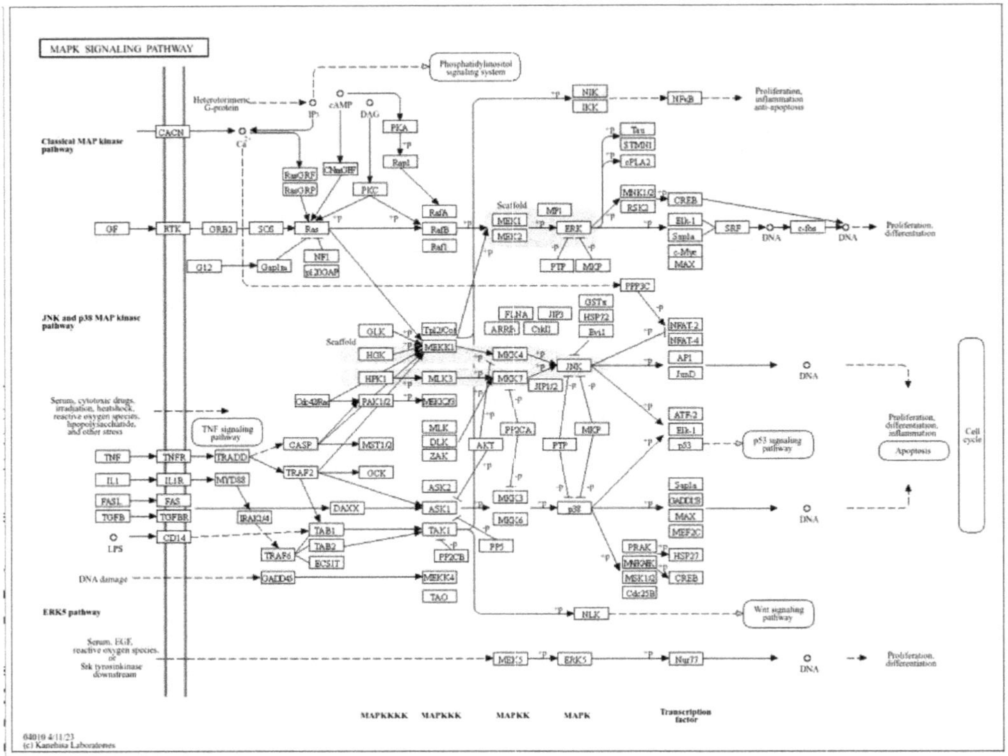

Figure 2.5: Diagram showing multiple inputs (EGF, cytokines, GPCR ligands) feeding into the MAPK cascade. Source: Genome. Link: https://www.genome.jp/pathway/map04010

2.3.5. Clinical Implications of Crosstalk

Understanding crosstalk is crucial in drug development, as inhibiting one pathway might unintentionally activate another, leading to resistance or adverse effects.

Cancer Therapy:

- Targeting **EGFR** alone may result in the activation of **alternate pathways** like PI3K-Akt or JAK-STAT.
- Combination therapies are developed to account for this.

Autoimmune and Inflammatory Disorders:
Drugs targeting **cytokine signaling** must consider how blocking one receptor might **amplify others** through feedback loops.

Summary

Crosstalk is an essential feature of the cellular signaling network that enables cells to respond flexibly and robustly to a complex environment. It ensures cellular behavior is context-dependent and not dictated by a single pathway in isolation.

2.4. Signal Amplification and Termination

Signal transduction in cells is not only about detection—it's about modulating the strength and duration of responses. Two crucial features that define the fidelity and flexibility of signaling pathways are amplification and termination. These processes ensure that the signal is strong enough to trigger a cellular response but also tightly regulated to avoid overstimulation or chronic activation.

2.4.1. Signal Amplification: Making the Message Louder

Amplification occurs when one signaling molecule activates multiple downstream targets, multiplying the effect of the original signal. This process is **non-linear** and can lead to massive cellular responses from minute quantities of ligand.

Key Amplification Mechanisms:

- **Enzymatic Cascades**
  Example: A single activated **G-protein** can activate **adenylyl cyclase**, which generates thousands of **cAMP molecules**, each of which activates **Protein Kinase A (PKA)** and triggers further downstream

events.

- **Second Messengers**
  Molecules like **cAMP**, **Ca^{2+}**, **IP_3**, and **DAG** act as internal amplifiers of signals.

- **Kinase Cascades**
  Example: The **MAPK cascade**, where each kinase activates multiple downstream kinases, resulting in exponential signal amplification.

2.4.2. Signal Termination: Turning the Message Off

Once a signal has been transmitted and cellular goals achieved, it must be turned off or attenuated. Without termination mechanisms, cells risk continuous stimulation, leading to diseases such as cancer, inflammation, or neurodegeneration.

Primary Termination Mechanisms:

Ligand Removal or Degradation
E.g., enzymatic degradation of extracellular ligands like acetylcholine by **acetylcholinesterase**.

Receptor Internalization and Downregulation
Activated receptors like **EGFR** may be internalized via **endocytosis**, then recycled or degraded.

GTP Hydrolysis in G-Proteins
GTP-bound G-proteins hydrolyze GTP to GDP, returning to an inactive state.

Phosphatases and Dephosphorylation
Enzymes like **protein phosphatase 2A (PP2A)** remove phosphate groups to inactivate signaling proteins.

Second Messenger Clearance

- cAMP is broken down by **phosphodiesterases (PDEs)**
- Ca^{2+} is pumped back into the **ER** or out of the cell

2.4.3. Balancing Amplification and Termination

Cells must balance signal amplification and termination dynamically:

Feedback Regulation

- **Negative feedback** (e.g., phosphatase activation) dampens signals.
- **Positive feedback** (e.g., Ca^{2+}-induced Ca^{2+} release) temporarily boosts signals but eventually fades.

Temporal Control

- Some signals, like hormone signaling, require **sustained activation**, while others, like neuronal synapses, require **rapid termination**.

In diseases like cancer, mutations often disrupt termination mechanisms (e.g., **constitutively active Ras or RTKs**), leading to persistent signal amplification and uncontrolled growth.

Summary

Signal amplification allows cells to respond sensitively to minimal stimuli, while termination mechanisms prevent overreaction and ensure homeostasis. A proper understanding of this balance is critical for **drug design** and **targeted therapies** in diseases where signaling becomes aberrant.

3. Cell Cycle and Cell Death Mechanisms

3.1. Regulation of the Cell Cycle

The **cell cycle** is a tightly regulated sequence of events through which a cell grows, replicates its DNA, and divides. This regulation is critical to ensure **genomic integrity**, proper cell function, and controlled proliferation. Dysregulation often leads to diseases such as **cancer**.

3.1.1. Overview of the Cell Cycle Phases

The eukaryotic cell cycle is divided into four main phases:

- **G_1 Phase (Gap 1):** Cell grows, produces organelles and proteins.
- **S Phase (Synthesis):** DNA replication occurs.
- **G_2 Phase (Gap 2):** Preparation for mitosis; more growth and error-checking.
- **M Phase (Mitosis):** Cell division (mitosis and cytokinesis).

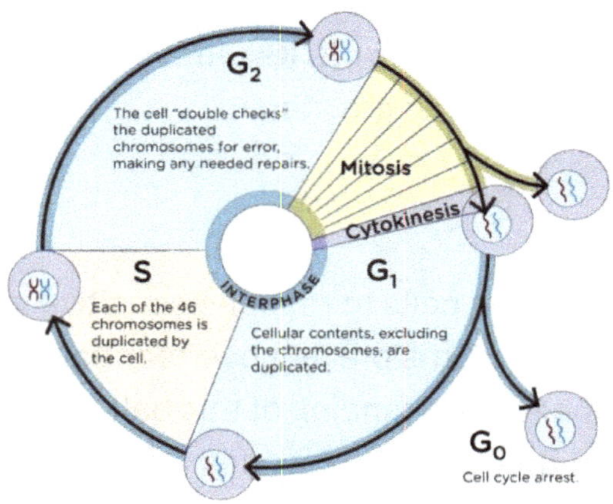

Figure 3.1: Diagram showing G_1, S, G_2, M, and G_0 phases in a circular representation. Source: Jack Westin. Link: https://jackwestin.com/resources/mcat-content/mitosis/phases-of-cell-cycle-g0-g1-s-g2-m

Additionally, cells may enter a **quiescent G_0 phase**, where they are metabolically active but not dividing.

3.1.2. Cyclins and Cyclin-Dependent Kinases (CDKs)

The core drivers of cell cycle progression are **cyclin-CDK complexes**. CDKs are serine/threonine kinases activated only when bound to specific **cyclins**, whose levels fluctuate throughout the cycle.

- **G_1/S Transition:** Cyclin D/CDK4, Cyclin E/CDK2
- **S Phase Entry:** Cyclin A/CDK2
- **G_2/M Transition:** Cyclin B/CDK1 (also known as **MPF**, mitosis-promoting factor)

These complexes phosphorylate target proteins that push the cell into the next phase.

3.1.3. Cell Cycle Checkpoints

Checkpoints are **control mechanisms** that assess cellular conditions before progression:

- **G_1 Checkpoint (Restriction Point):** Checks DNA damage, growth factors
- **G_2 Checkpoint:** Checks DNA replication accuracy and cell size
- **M Checkpoint (Spindle Assembly):** Ensures proper chromosome alignment

Checkpoint regulators include:

- **p53:** Halts cycle in response to DNA damage; activates p21 to inhibit CDKs

- **ATM/ATR kinases:** Sense DNA damage and initiate repair pathways
- **MAD/BUB proteins:** Monitor spindle integrity in metaphase

3.1.4. Regulation by External Signals

The cell cycle can be influenced by:

- **Mitogens (e.g., EGF, PDGF):** Promote cyclin expression and progression
- **Growth Inhibitors (e.g., TGF-β):** Activate CDK inhibitors like **p15, p16**
- **Contact Inhibition:** Cell-cell contact can arrest cycle in G_1
- **Nutrient Availability:** Scarcity halts cells in G_0 or G_1

3.1.5. Dysregulation and Disease

Disruption in any component—cyclins, CDKs, checkpoints—can result in uncontrolled proliferation. Many cancers involve:

- Overexpression of **cyclin D1**
- Loss-of-function mutations in **p53**
- CDK hyperactivity due to loss of CDK inhibitors

Summary

The cell cycle is a masterfully orchestrated process dependent on cyclins, CDKs, and regulatory checkpoints. Understanding its regulation is fundamental to modern cell biology and critical in the context of cancer biology and therapeutics.

3.2. Checkpoints and Cyclin-Dependent Kinases

The cell cycle is governed by a series of checkpoints that ensure the fidelity of DNA replication, division, and overall cell health. At the heart of this regulation are **cyclin-dependent kinases (CDKs)**, whose activity is tightly controlled by cyclins, phosphorylation, and inhibitors. Checkpoints act as surveillance mechanisms that halt cell cycle progression until the cell meets specific criteria, thereby safeguarding genomic integrity.

3.2.1. Overview of Cell Cycle Checkpoints

The three major checkpoints are:

- **G_1/S Checkpoint (Restriction Point):** Verifies cell size, DNA integrity, and presence of growth factors.
- **G_2/M Checkpoint:** Confirms successful DNA replication and checks for DNA damage.
- **Spindle Assembly Checkpoint (Metaphase-Anaphase):** Ensures that all chromosomes are properly attached to spindle microtubules before chromosome segregation.

These checkpoints prevent the cell from moving forward with division when abnormalities are detected.

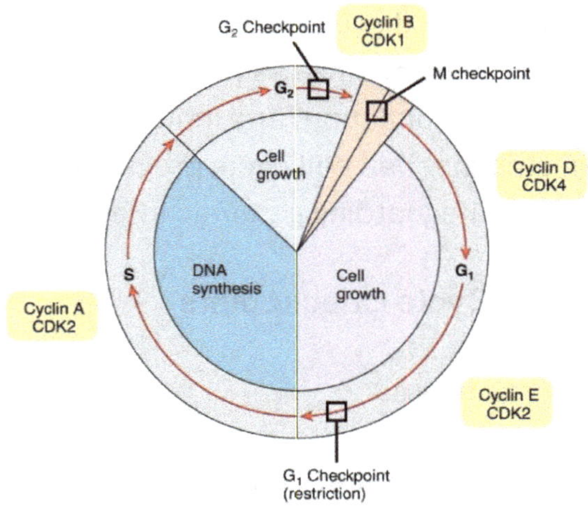

Figure 3.2: Diagram showing the G_1/S, G_2/M, and spindle checkpoints along a circular cell cycle path. Source: Wikipedia. Link: https://en.wikipedia.org/wiki/Cell_cycle_checkpoint#/media/File:0332_Cell_Cycle_With_Cyclins_and_Checkpoints.jpg

3.2.2. Cyclins and CDK Complexes

Cyclins are regulatory proteins whose concentrations rise and fall cyclically throughout the cell cycle. When bound to **CDKs**, they activate the kinase's ability to phosphorylate target proteins that drive the cell into the next phase of the cycle.

Cyclin	CDK Partner	Cell Cycle Phase
Cyclin D	CDK4/6	G_1 phase
Cyclin E	CDK2	G_1/S transition
Cyclin A	CDK2	S phase
Cyclin B	CDK1	G_2/M transition

These complexes are tightly regulated by synthesis/degradation of cyclins, phosphorylation states of CDKs, and the presence of **CDK inhibitors**.

3.2.3. CDK Activation and Inhibition

Activation requires:

- Binding to a cyclin
- Phosphorylation at activating sites (e.g., Thr160 on CDK2)
- Removal of inhibitory phosphates by **Cdc25 phosphatases**

Inhibition occurs through:

1. Phosphorylation at inhibitory sites (e.g., Tyr15)

2. Binding to CDK inhibitors (CKIs):

- **INK4 family (p15, p16):** Block CDK4/6
- **Cip/Kip family (p21, p27):** Inhibit CDK2, CDK1

Checkpoint signals, especially DNA damage, activate tumor suppressors like **p53**, which upregulate **p21**, halting the cell cycle.

3.2.4. The Role of p53 in Checkpoint Control

p53 is a critical tumor suppressor protein and transcription factor that becomes activated in response to DNA damage or stress. It induces the transcription of **p21**, which inhibits CDK2/cyclin E and CDK1/cyclin B complexes, thus halting cell cycle progression at G_1/S and G_2/M checkpoints.

Loss of p53 function is a common hallmark in many cancers and leads to defective checkpoint responses and uncontrolled proliferation.

3.2.4. The Role of p53 in Checkpoint Control

p53 is a critical tumor suppressor protein and transcription factor that becomes activated in response to DNA damage or stress. It induces the transcription of **p21**, which inhibits CDK2/cyclin E and CDK1/cyclin B complexes, thus halting cell cycle progression at G_1/S and G_2/M checkpoints.

Loss of p53 function is a common hallmark in many cancers and leads to defective checkpoint responses and uncontrolled proliferation.

3.2.5. Checkpoint Dysregulation in Cancer

Cancer cells often exhibit:

- **Loss of CDK inhibitors** (e.g., mutations in **p16**, **p21**)
- **Overexpression of cyclins**, especially Cyclin D1
- **Loss of checkpoint proteins**, such as **Rb** or **p53**

Such alterations bypass normal cell cycle control, leading to unchecked proliferation and genomic instability.

Summary

The coordination between **checkpoints**, **cyclins**, and **CDKs** forms the backbone of cell cycle regulation. Disruption of this coordination—whether through loss of checkpoint integrity or overactivation of CDKs—is a major driver of oncogenesis and a target for therapeutic intervention.

3.3. Apoptosis, Autophagy, and Necroptosis

Understanding the pathways of programmed cell death is fundamental to comprehending how cells maintain homeostasis, respond to stress, and defend against disease. This section explores three major forms of cell death: **apoptosis**, **autophagy**, and **necroptosis**, each with distinct molecular mechanisms and cellular consequences.

3.3.1. Apoptosis: Programmed Cell Death

Apoptosis is a highly regulated process that leads to cell death without provoking an inflammatory response. It is crucial in development, immune regulation, and the removal of damaged cells.

Key Features:

- Cell shrinkage and chromatin condensation
- DNA fragmentation
- Membrane blebbing and apoptotic bodies

Apoptosis is mediated via two main pathways:

- **Intrinsic (mitochondrial) pathway**
- **Extrinsic (death receptor) pathway**

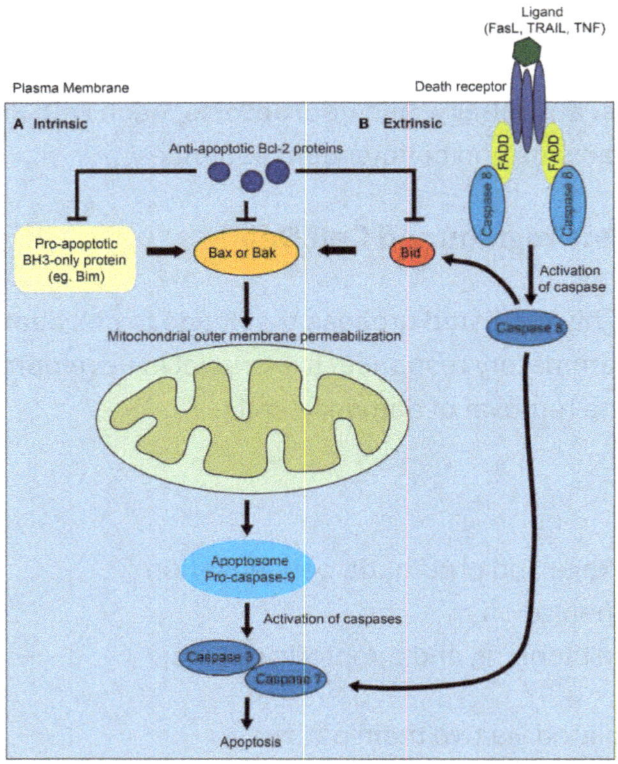

Figure 3.3: Diagram showing intrinsic and extrinsic apoptotic pathways. Source: ResearchGate. Link: https://www.researchgate.net/figure/Diagram-of-intrinsic-and-extrinsic-pathways-of-apoptosis-A-In-the-intrinsic-pathway_fig1_269181128

Important Proteins:

- Caspases (initiator: caspase-8, -9; effector: caspase-3)
- Bcl-2 family proteins (pro-apoptotic: Bax, Bak; anti-apoptotic: Bcl-2, Bcl-xL)
- Cytochrome c, Apaf-1 (intrinsic)

3.3.2. Autophagy: Cellular Recycling System

Autophagy is a survival mechanism activated during starvation, stress, or damaged organelles. It allows the cell to degrade and recycle its own components through the lysosomal pathway.

Types of Autophagy:

- **Macroautophagy** (most common)
- Microautophagy
- Chaperone-mediated autophagy

Stages of Autophagy:

1. Initiation
2. Autophagosome formation
3. Fusion with lysosome
4. Degradation

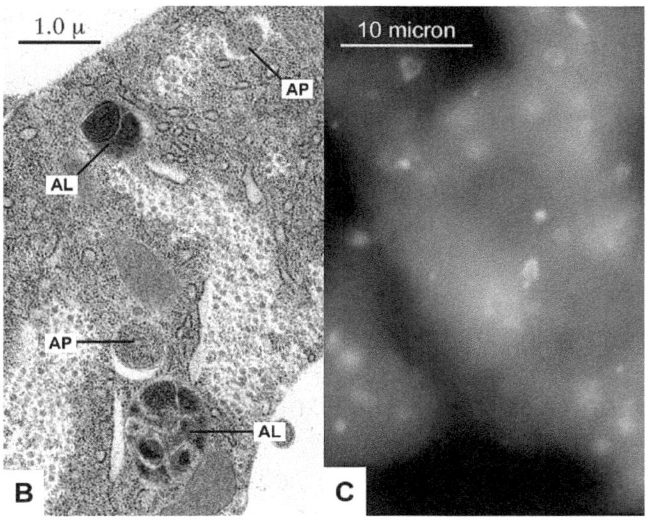

Figure 3.4: Autophagy Pathway Diagram.
Source: Wikimedia. Link: https://commons.wikimedia.org/wiki/File:Autophagy.png

Key Regulators:

- mTOR (inhibits autophagy)
- AMPK (activates autophagy)
- Beclin-1, LC3-II

Autophagy can be cytoprotective but excessive autophagy may lead to autophagic cell death.

3.3.3. Necroptosis: Regulated Necrosis

Necroptosis is a **programmed form of necrosis**, distinct from apoptosis, that results in **cell membrane rupture and inflammation**. It is often a backup mechanism when apoptosis is inhibited.

Activation Pathway:

- Triggered by death receptors (e.g., TNFR1)
- Requires **RIPK1**, **RIPK3**, and **MLKL**

Necroptosis plays roles in:

- Inflammation
- Neurodegeneration
- Viral infection responses

3.3.4. Interplay and Cross-Regulation Between Pathways

These cell death pathways are not isolated but **interconnected**. For example:

- Caspase-8 inhibits necroptosis by cleaving RIPK1.

- Bcl-2 inhibits both apoptosis and autophagy.
- Certain stress signals can switch the cell from autophagy to apoptosis.

Clinical Relevance:

- Dysregulation of these processes is linked to **cancer**, **autoimmune diseases**, **neurodegenerative disorders**, and **infection**.
- Therapeutic strategies aim to modulate these pathways using small molecules, gene therapy, or immunomodulation.

3.4. Implications in Disease and Cancer

Understanding the molecular mechanisms governing the cell cycle and cell death pathways is critical in exploring how their dysregulation contributes to disease, especially cancer. Abnormalities in checkpoints, overexpression of cyclins, evasion of apoptosis, and defective DNA repair mechanisms can lead to unchecked cell proliferation and genomic instability. This section delves into how these processes underpin various diseases, with an emphasis on cancer biology.

3.4.1. Oncogenes and Tumor Suppressors

Oncogenes are mutated or overexpressed versions of normal proto-oncogenes that drive excessive cell proliferation. In contrast, tumor suppressor genes inhibit the cell cycle or promote apoptosis. Loss-of-function mutations in tumor suppressors like *TP53*, *RB1*, or *BRCA1/2* are commonly found in cancers.

3.4.2. Apoptosis Evasion and Cancer Progression

Cancer cells often develop mechanisms to evade apoptosis, contributing to their survival. Common strategies include overexpression of anti-apoptotic proteins (e.g., Bcl-2) or downregulation of pro-apoptotic factors (e.g., Bax). Mutations in *p53*, a key regulator of apoptosis, are found in over 50% of human cancers.

3.4.3. Uncontrolled Proliferation and Checkpoint Failure

Checkpoint proteins like ATM, ATR, CHK1, and CHK2 help maintain genomic integrity. When these are mutated or inactivated, cells bypass DNA damage checkpoints, leading to the accumulation of mutations and chromosomal instability.

3.4.4. Role in Other Diseases Beyond Cancer

Besides cancer, deregulated cell cycle and death mechanisms are implicated in neurodegenerative disorders (e.g., Alzheimer's, Parkinson's), autoimmune diseases, and developmental defects. Aberrant autophagy can promote either cell survival or cell death, depending on the context.

3.4.5. Therapeutic Interventions Targeting These Pathways

Understanding how these pathways are altered in disease has led to targeted therapies. CDK inhibitors (e.g., Palbociclib), Bcl-2 inhibitors (e.g., Venetoclax), and PARP inhibitors (e.g., Olaparib for BRCA-mutated cancers) are key examples.

Part II: Molecular Biology at an Advanced Level

4. DNA Structure, Chromatin, and Epigenetics

4.1. DNA Topology and Supercoiling

DNA topology refers to the spatial arrangement of the DNA double helix, which can be altered by twisting, bending, or coiling. In cellular environments, DNA is rarely found in a relaxed B-form helix; instead, it experiences supercoiling, which plays a critical role in processes such as replication, transcription, and chromatin organization.

4.1.1. Basic DNA Topological States

DNA exists in different topological states based on its twist and writhe:

- **Relaxed DNA**: The double helix in its most stable B-form.
- **Positive supercoiling**: Overwinding of the helix, making it more compact and energetically strained.
- **Negative supercoiling**: Underwinding, which facilitates strand separation during transcription and replication.

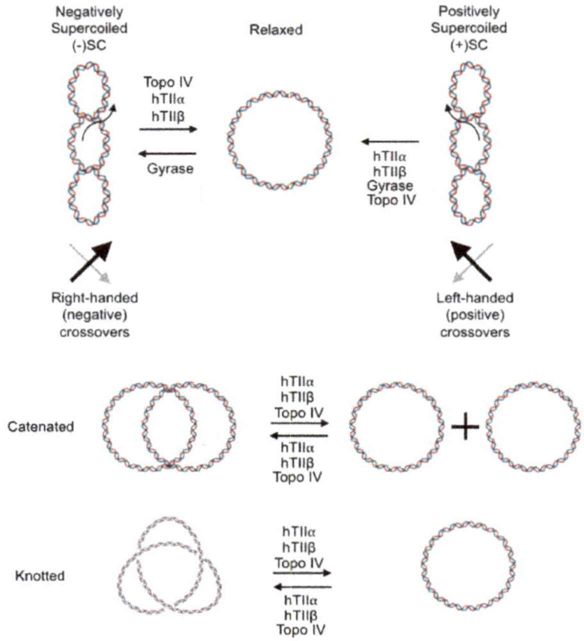

Figure 4.1: Representation of relaxed, positively supercoiled, and negatively supercoiled DNA. Source: MDPI. Link: https://www.mdpi.com/1422-0067/24/13/11199

4.1.2. Role of Topoisomerases

Topoisomerases are enzymes that regulate DNA topology by making transient cuts in the DNA backbone to relieve or introduce supercoils:

- **Topoisomerase I**: Cuts one strand to relieve torsional stress.
- **Topoisomerase II (including DNA gyrase in bacteria)**: Cuts both strands, allowing for decatenation and untangling.

These enzymes are critical during DNA replication, transcription, and chromosome segregation.

4.1.3. DNA Supercoiling in Chromatin Context

In eukaryotic cells, supercoiling works in concert with chromatin packaging:

- Nucleosome formation can constrain supercoiling.
- Chromatin remodeling enzymes help manage torsional strain during transcription.
- Supercoiling is regulated regionally, depending on transcriptional activity.

4.1.4. Implications for Gene Expression and Replication

Supercoiling influences:

- **Promoter accessibility**: Negative supercoiling enhances strand separation at promoters.
- **Transcription elongation**: RNA polymerase generates positive supercoils ahead and negative ones behind it.
- **Replication fork progression**: Supercoiling must be resolved for fork movement and stability.

Disruption of normal supercoiling is linked to genome instability and can be a target for antimicrobial and anticancer drugs.

4.2. Chromatin Remodeling and Histone Modifications

Chromatin remodeling and histone modifications are essential mechanisms that regulate DNA accessibility and gene expression. These processes influence how tightly DNA is wrapped around histones and how the chromatin landscape is organized across the genome, thereby controlling the functional output of the genome in different cell types and states.

4.2.1. Chromatin Structure and Accessibility

Chromatin exists in two main forms:

- **Euchromatin**: Loosely packed, transcriptionally active.
- **Heterochromatin**: Densely packed, transcriptionally silent.

DNA is wrapped around histone octamers to form nucleosomes, which are the fundamental repeating units of chromatin. Accessibility of DNA to transcription factors and polymerases depends on how tightly it is packaged.

4.2.2. Chromatin Remodeling Complexes

ATP-dependent chromatin remodeling complexes alter nucleosome positioning, spacing, and composition. Major families include:

- **SWI/SNF**: Disrupts nucleosomes to promote transcription.
- **ISWI**: Maintains nucleosome spacing.
- **CHD**: Involved in both repression and activation.
- **INO80**: Facilitates DNA repair and replication.

These complexes reposition or evict histones, altering chromatin compaction and gene accessibility.

4.2.3. Histone Modifications: The Histone Code

Histone proteins undergo post-translational modifications (PTMs), primarily on their N-terminal tails. Common modifications include:

- **Acetylation**: Typically associated with gene activation.
- **Methylation**: Can activate or repress depending on the residue.

- **Phosphorylation**: Involved in DNA repair and chromatin condensation.
- **Ubiquitination**: Plays roles in transcriptional regulation and histone turnover.

These modifications collectively form a "histone code" interpreted by chromatin-binding proteins.

4.2.4. Writers, Erasers, and Readers

Histone modifications are regulated by:

- **Writers**: Enzymes that add modifications (e.g., HATs, HMTs).
- **Erasers**: Enzymes that remove modifications (e.g., HDACs, demethylases).
- **Readers**: Proteins that recognize and bind specific modifications (e.g., bromodomains, chromodomains).

These regulators dynamically control gene expression and chromatin behavior in response to cellular signals.

4.2.5. Role in Development and Disease

Aberrant histone modifications and remodeling activity are implicated in:

- Developmental disorders
- Cancer progression
- Neurological diseases

Targeting epigenetic enzymes (like HDAC inhibitors) has emerged as a promising therapeutic approach.

4.3. DNA Methylation and Non-coding RNAs

DNA methylation and non-coding RNAs (ncRNAs) represent two powerful and interconnected epigenetic regulatory mechanisms that control gene expression, genome stability, and developmental programming. Together, they shape the epigenome and influence cell fate, identity, and disease susceptibility.

4.3.1. DNA Methylation: Mechanism and Enzymes

DNA methylation involves the addition of a methyl group ($-CH_3$) to the 5-carbon of cytosine, primarily in CpG dinucleotides. This modification:

- Silences gene expression
- Maintains genomic stability
- Contributes to X-chromosome inactivation and imprinting

Key enzymes involved:

- **DNMT1**: Maintains methylation patterns during DNA replication.
- **DNMT3A and DNMT3B**: Establish new methylation marks (de novo methylation).

4.3.2. CpG Islands and Methylation Patterns

CpG islands are regions rich in CpG dinucleotides, often found near gene promoters. In normal cells:

- Active genes have **unmethylated CpG islands**.
- Silenced genes or imprinted alleles may have **methylated promoters**.

Aberrant methylation of CpG islands can silence tumor suppressor genes in cancer or alter developmental gene expression.

4.3.3. Non-coding RNAs: Classification and Functions

Non-coding RNAs (ncRNAs) are RNA molecules that do not encode proteins but regulate gene expression at transcriptional and post-transcriptional levels. They are broadly categorized into:

- **Short ncRNAs** (e.g., miRNA, siRNA, piRNA)
- **Long ncRNAs** (lncRNAs, >200 nucleotides)

Their functions include:

- Gene silencing
- Chromatin modification
- Transcriptional interference
- Post-transcriptional mRNA regulation

4.3.4. MicroRNAs (miRNAs) and Small Interfering RNAs (siRNAs)

- **miRNAs**: Endogenously expressed; regulate gene expression by binding 3' UTR of target mRNAs, causing degradation or translation inhibition.
- **siRNAs**: Often exogenous or derived from long dsRNA; trigger specific mRNA cleavage via RISC complex.

These small RNAs act as fine-tuners of gene expression and play roles in development, immune responses, and disease.

4.3.5. Long Non-coding RNAs (lncRNAs) and Epigenetic Regulation

lncRNAs regulate gene expression by:

- Recruiting chromatin modifiers (e.g., PRC2)
- Acting as scaffolds, decoys, or guides
- Influencing nuclear architecture and mRNA processing

A notable example is **XIST**, which coats the inactive X chromosome and triggers its silencing via histone modifications and DNA methylation.

4.3.6. Interplay Between DNA Methylation and ncRNAs

DNA methylation and non-coding RNAs interact in complex feedback loops:

- Methylation patterns can regulate ncRNA expression.
- ncRNAs can recruit DNA methyltransferases or demethylases to specific loci.

These interactions form intricate epigenetic networks critical for development, cell differentiation, and cancer biology.

4.4. Epigenetic Regulation in Development and Disease

Epigenetic modifications such as DNA methylation, histone modifications, and non-coding RNAs orchestrate gene expression programs crucial for embryonic development, cell differentiation, and tissue homeostasis. Disruption of these processes can lead to developmental disorders, cancer, neurological diseases, and aging-related conditions.

4.4.1. Epigenetics in Embryonic Development

During development, pluripotent stem cells undergo progressive epigenetic reprogramming to establish lineage-specific gene expression profiles. Key processes include:

- **Global DNA demethylation and remethylation** during early embryogenesis.
- **Establishment of tissue-specific enhancers** and chromatin landscapes.
- **Silencing of pluripotency genes** and activation of lineage-defining transcription factors.

4.4.2. Genomic Imprinting and Epigenetic Memory

Genomic imprinting is an epigenetic phenomenon where only one parental allele is expressed while the other is silenced. This is mediated by:

- **Parent-specific DNA methylation patterns** established in gametes.
- **Histone modifications** that reinforce silencing.

Imprinted genes are essential for growth, development, and neurological function. Errors in imprinting cause disorders such as **Prader-Willi** and **Angelman syndromes**.

4.4.3. Epigenetic Dysregulation in Cancer

Cancer cells exhibit widespread epigenetic alterations, including:

- **Hypermethylation** of tumor suppressor gene promoters (e.g., p16, BRCA1)
- **Global hypomethylation**, leading to genomic instability
- **Aberrant histone modification patterns**

- **Misregulation of ncRNAs**

These changes can drive uncontrolled proliferation, immune evasion, and metastasis.

4.4.4. Epigenetics in Neurological and Psychiatric Disorders

Epigenetic modifications are implicated in brain development, learning, and memory. Disruptions can lead to:

- **Neurodevelopmental disorders** (e.g., Rett syndrome, Fragile X)
- **Neurodegenerative diseases** (e.g., Alzheimer's, Parkinson's)
- **Psychiatric conditions** (e.g., schizophrenia, depression)

Environmental factors like stress, toxins, and diet can modify the brain epigenome across the lifespan.

4.4.5. Epigenetic Contributions to Aging and Longevity

Aging is accompanied by global and site-specific changes in DNA methylation ("epigenetic clock"), histone modification shifts, and altered ncRNA profiles. Key features include:

- **Loss of heterochromatin integrity**
- **Increase in DNA methylation drift**
- **Altered expression of aging-related genes (e.g., SIRT1, FOXO)**

These contribute to cellular senescence, inflammation, and functional decline.

5. Advanced Gene Expression and Regulation

5.1. Transcriptional Control Mechanisms

Transcriptional control is the first and most regulated step in gene expression. It determines when, where, and how much of a gene is transcribed into RNA, enabling cells to respond to internal and external signals. This regulation is achieved through complex interactions among transcription factors, promoter elements, enhancers, repressors, chromatin structure, and epigenetic marks.

5.1.1. Core Promoters and Transcription Factors

Core promoters are DNA sequences located upstream of the transcription start site (TSS), where general transcription factors and RNA polymerase II assemble to begin transcription. Key elements include the TATA box, initiator (Inr), and downstream promoter elements (DPEs).

General Transcription Factors (GTFs) such as TFIID, TFIIA, and TFIIB are essential for RNA polymerase II binding. In contrast, **specific transcription factors** bind to regulatory sequences to activate or repress gene expression.

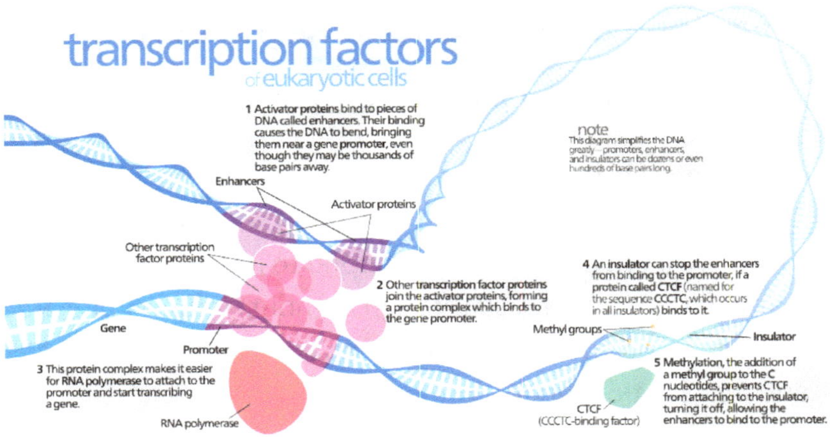

Figure 5.1: Binding of general and specific transcription factors to the promoter region. Source: Wikipedia. Link: https://en.wikipedia.org/wiki/General_transcription_factor

5.1.2. Enhancers, Silencers, and Insulators

- **Enhancers** are distal DNA elements that increase transcription efficiency through looping interactions with promoters.
- **Silencers** are similar in structure but repress gene transcription.
- **Insulators** block interaction between enhancers and promoters when positioned in between them.

These elements can act over long distances and are often tissue-specific.

5.1.3. Chromatin Structure and Transcriptional Accessibility

Gene accessibility is governed by the local chromatin environment:

- **Euchromatin** is loosely packed and transcriptionally active.
- **Heterochromatin** is condensed and generally silenced.
- **Nucleosome remodeling complexes** such as SWI/SNF reposition

nucleosomes to expose regulatory regions.

- **Histone acetylation** (e.g., H3K27ac) and methylation marks modulate transcriptional readiness.

5.1.4. Combinatorial and Cooperative Regulation

Genes are often controlled by multiple transcription factors acting in combination, allowing fine-tuned regulation. These factors may:

- **Cooperate synergistically** to enhance gene activation.
- **Compete** for overlapping binding sites.
- **Interact with cofactors or corepressors** to influence chromatin and transcription machinery.

This combinatorial logic underlies the complexity of gene expression in development, immunity, and disease.

5.1.5. Response to Extracellular Signals

Signal transduction cascades (e.g., MAPK, JAK-STAT, NF-κB) often culminate in the activation or repression of transcription factors. These factors then bind DNA to regulate expression of specific genes in response to hormones, stress, or growth factors.

Examples:

- **Glucocorticoid receptor** binding in response to cortisol.
- **STAT proteins** activation by cytokines.
- **CREB phosphorylation** after cAMP signaling.

5.2. RNA Processing and RNA Editing

RNA processing and RNA editing are crucial post-transcriptional mechanisms that regulate the diversity, stability, and functionality of RNA molecules. These steps transform the primary RNA transcript (pre-mRNA) into a mature mRNA capable of being translated into proteins. The processing includes capping, splicing, and polyadenylation, while editing involves direct alterations of the RNA sequence, influencing gene expression without modifying the underlying DNA.

5.2.1. Pre-mRNA Processing in Eukaryotes

In eukaryotic cells, the primary RNA transcript undergoes several modifications before it can function as mature mRNA:

- **5' Capping:** A 7-methylguanosine cap is added to the 5' end shortly after transcription begins. This cap protects the mRNA from degradation and is essential for ribosome recognition.
- **Splicing:** Introns (non-coding regions) are removed, and exons (coding regions) are joined together by the spliceosome. Alternative splicing allows for different proteins to be produced from a single gene.
- **3' Polyadenylation:** A poly(A) tail is added to the 3' end, enhancing the stability and translation efficiency of the mRNA.

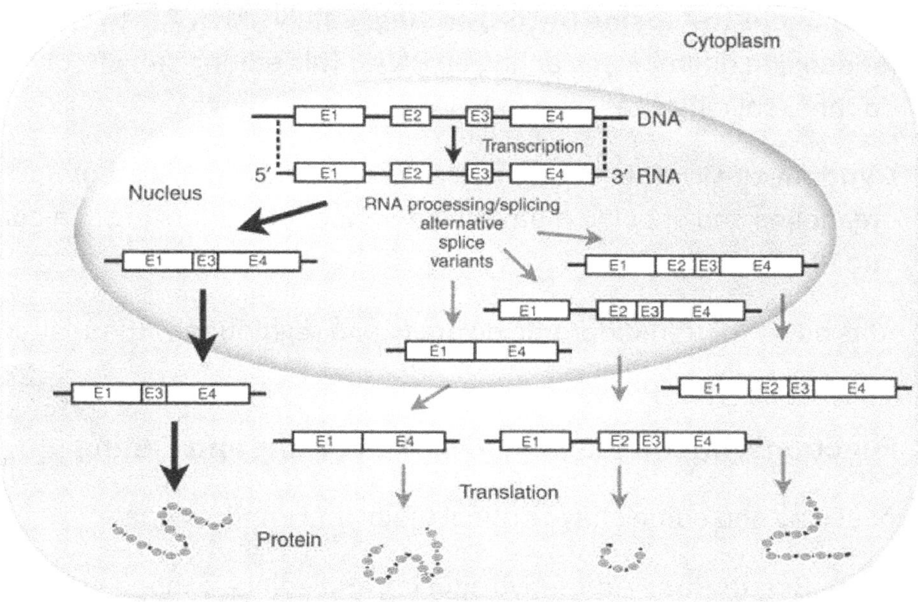

Figure 5.2: A schematic illustration of transcription, pre-mRNA processing/splicing and translation. Source: ResearchGate. Link: https://www.researchgate.net/figure/Schematic-illustration-of-transcription-pre-mRNA-processing-splicing-and-translation-A_fig1_23498088

5.2.2. RNA Splicing and Spliceosome Dynamics

The spliceosome, a large ribonucleoprotein complex, orchestrates the removal of introns. It recognizes specific sequences at the intron-exon boundaries and catalyzes two transesterification reactions to excise introns and ligate exons.

Alternative splicing expands proteomic diversity and allows a single gene to encode multiple protein variants. Regulatory sequences and splicing factors determine which exons are included or skipped.

5.2.3. RNA Editing: A-to-I and C-to-U Conversions

RNA editing involves chemical modifications that change nucleotide sequences after transcription. The two most well-studied types are:

- **Adenosine-to-Inosine (A-to-I) editing**, catalyzed by ADAR enzymes (adenosine deaminases acting on RNA). This editing can affect codons, splicing sites, and miRNA binding.
- **Cytidine-to-Uridine (C-to-U) editing**, primarily seen in plant organelles and specific mammalian transcripts, alters protein coding and RNA stability.

These edits allow for transcriptome diversity and regulation without altering the DNA.

5.2.4. Functional Implications of RNA Processing and Editing

RNA processing and editing have critical biological implications:

- **Gene Regulation:** Alternative splicing and editing expand protein diversity.
- **Development:** Tissue-specific splicing isoforms are vital during differentiation.
- **Disease Associations:** Mutations in splicing factors or aberrant editing patterns are linked to neurological disorders and cancers.

5.3. Translational Control and Riboswitches

Translational control plays a critical role in fine-tuning gene expression by regulating how and when mRNAs are translated into proteins. This regulation ensures cellular adaptation to environmental cues, developmental signals, and stress conditions. Riboswitches, structured RNA elements, are one such mechanism, directly sensing metabolites and modulating gene expression at the translational level.

5.3.1. Mechanisms of Translational Control

Translational control occurs at various stages:

- **Initiation Regulation:** The most common control point. Eukaryotic initiation factors (eIFs) mediate the recruitment of ribosomes. Phosphorylation of eIF2α or sequestering of eIF4E can inhibit translation under stress or during viral infection.
- **mRNA Localization:** Certain mRNAs are transported to specific cellular locations and translated locally. This is important in neurons and during embryonic development.
- **mRNA Stability and Degradation:** mRNAs with short poly(A) tails or AU-rich elements are rapidly degraded, limiting translation.
- **Upstream Open Reading Frames (uORFs):** Regulatory sequences in the 5' UTR can prevent the main ORF from being translated.

5.3.2. Role of microRNAs in Translational Repression

MicroRNAs (miRNAs) are small non-coding RNAs that bind to complementary sequences in the 3' UTR of target mRNAs. This binding can block translation or promote mRNA degradation via the RNA-induced silencing complex (RISC).

Mechanism: miRNAs repress translation by:

- Preventing ribosome assembly.
- Inducing deadenylation and decapping.
- Recruiting proteins that promote mRNA decay.

This regulation is widespread in animals and plants and plays roles in development, immunity, and cancer.

5.3.3. Riboswitches: RNA-Based Metabolite Sensors

Riboswitches are structured RNA domains typically located in the 5' untranslated region (UTR) of bacterial mRNAs. They bind metabolites or ions and undergo conformational changes that affect gene expression.

Ligand Binding: Upon ligand binding, the riboswitch may:

- Mask or reveal the ribosome binding site (RBS).
- Induce premature transcription termination.
- Alter mRNA stability.

Examples:

- **Thiamine pyrophosphate (TPP) riboswitch**: regulates genes involved in vitamin B1 synthesis.
- **FMN riboswitch**: controls flavin biosynthesis.

5.3.4. Translational Control in Development and Stress Responses

In multicellular organisms, translational regulation is crucial during:

- **Embryogenesis:** Maternal mRNAs are stored and selectively translated during early development.

- **Stress Response:** Under heat shock or nutrient starvation, cells suppress global translation while selectively translating stress-response proteins (e.g., heat shock proteins).

5.4. Gene Silencing: RNAi, miRNA, and siRNA

Gene silencing refers to the regulation of gene expression in a way that prevents the production of specific proteins. One of the most powerful gene silencing mechanisms is **RNA interference (RNAi)**, which is a post-transcriptional gene regulation system involving small RNA molecules—

primarily **microRNAs (miRNAs)** and **small interfering RNAs (siRNAs).** These molecules play crucial roles in development, immune defense, and experimental gene knockdown.

5.4.1. Mechanisms of RNA Interference (RNAi)

RNAi is a natural biological process in which **double-stranded RNA (dsRNA)** triggers the degradation or translational repression of complementary mRNA sequences.

- **Initiation:** dsRNA is introduced or produced in the cell.
- **Dicer Enzyme:** Cleaves dsRNA into ~21-23 nucleotide siRNAs.
- **RISC Complex:** The RNA-induced silencing complex unwinds siRNA and incorporates one strand as a guide.
- **mRNA Targeting:** The guide RNA leads RISC to complementary mRNA, which is either cleaved (siRNA) or translationally repressed (miRNA).

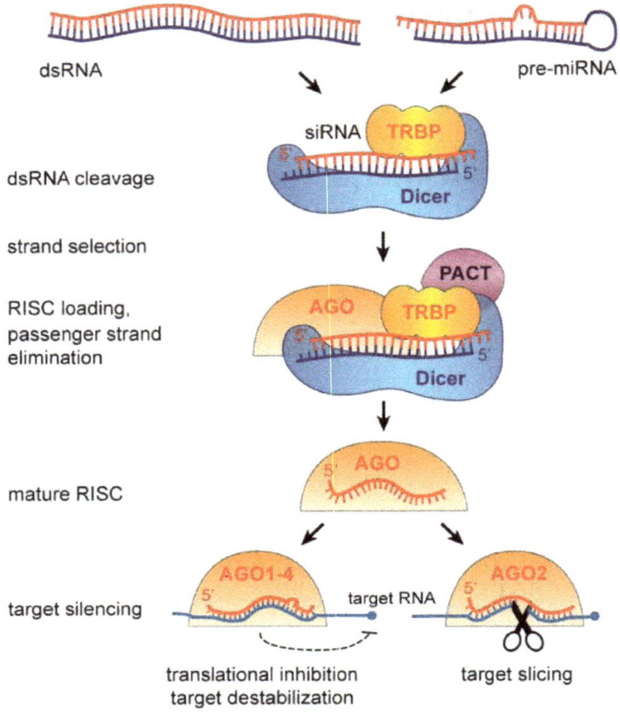

Figure 5.3: The RNA interference pathway showing the roles of Dicer and the RISC complex. Source: ResearchGate: Link: https://www.researchgate.net/figure/The-RNA-interference-RNAi-pathway-in-mammals-A-single-Dicer-protein-processes-long_fig4_333158892

5.4.2. microRNAs (miRNAs)

miRNAs are endogenous, non-coding RNAs transcribed from miRNA genes. They fold into hairpin loops, are processed by Dicer, and associate with RISC.

- **Target Recognition:** Partial complementarity to target mRNAs, usually in the 3' untranslated region (3' UTR).
- **Function:** Inhibit translation or lead to mRNA destabilization.
- **Biological Importance:** miRNAs are involved in cellular

differentiation, development, metabolism, and disease regulation.

5.4.3. small interfering RNAs (siRNAs)

siRNAs are usually exogenous (introduced by viruses or researchers) but can also originate from endogenous sources.

- **Target Recognition:** Perfect or near-perfect complementarity to target mRNA.
- **Function:** Guide RISC to cleave and degrade specific mRNAs.
- **Research Tool:** Widely used in genetic research to knock down gene expression selectively.

5.4.4. Applications of Gene Silencing

Gene silencing technologies have widespread applications:

- **Functional Genomics:** Silencing genes to study their functions.
- **Therapeutics:** siRNA-based drugs are in development for cancer, viral infections, and genetic disorders.
- **Agriculture:** RNAi used to develop pest-resistant or drought-tolerant crops.
- **Antiviral Defense:** Many organisms use RNAi to combat viral RNA.

6. Genome Engineering and CRISPR Technologies

6.1. CRISPR-Cas Systems

The **CRISPR-Cas system** (Clustered Regularly Interspaced Short Palindromic Repeats and CRISPR-associated proteins) is a revolutionary tool for **genome editing**, allowing researchers to alter DNA sequences and modify gene function with unprecedented precision. Originally discovered as a bacterial immune defense against viruses, CRISPR has been adapted into one of the most powerful and versatile tools in molecular biology.

6.1.1. Discovery and Biological Role of CRISPR

CRISPR was first identified in bacteria and archaea as a defense mechanism against bacteriophages. The system captures snippets of viral DNA and integrates them into the host genome at the CRISPR loci. These segments serve as a genetic memory to recognize and defend against future infections.

Key Players:

- **CRISPR Array:** Repeats and spacers derived from viral DNA.
- **Cas Proteins:** Enzymes (e.g., Cas9, Cas12a) that use CRISPR RNA (crRNA) to locate and cut complementary viral DNA.

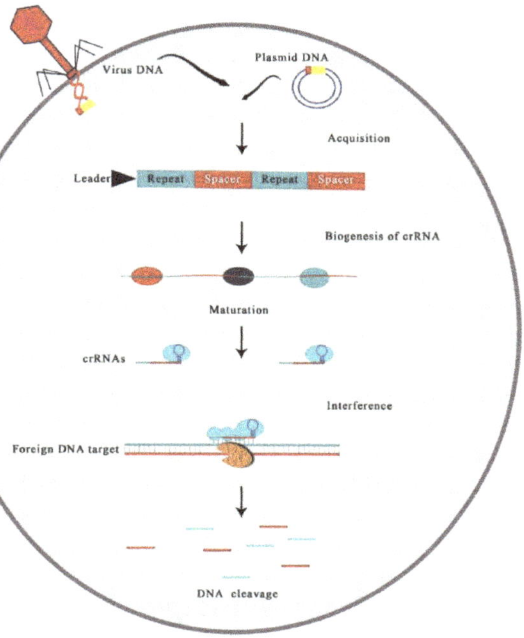

Figure 6.1: Natural bacterial CRISPR defense showing spacer acquisition, crRNA biogenesis, and interference with foreign DNA. *Source: ResearchGate: Link: https://www.researchgate.net/figure/Mechanism-of-CRISPR-cas-immunity-divided-into-three-stages-Stage-1-spacer-acquisition_fig1_308611937*

6.1.2. CRISPR-Cas9: Mechanism of Genome Editing

In genome editing, the CRISPR-Cas9 system is engineered to target specific genomic sequences.

- **Guide RNA (gRNA):** Designed to match the target DNA sequence.
- **Cas9 Nuclease:** Binds gRNA and creates a double-stranded break (DSB) in the DNA.
- **Cellular Repair:**
    - **Non-Homologous End Joining (NHEJ):** Error-prone, often results in gene knockout.
    - **Homology-Directed Repair (HDR):** Precise editing using a

donor DNA template.

6.1.3. Types of CRISPR-Cas Systems

There are multiple types of CRISPR systems, but **Class 2 systems** (especially Cas9 and Cas12a) are most commonly used in genome editing.

- **Cas9 (Type II):** Requires a single guide RNA (sgRNA); creates blunt-ended cuts.
- **Cas12a (Cpf1, Type V):** Processes its own crRNA and creates staggered cuts.
- **Cas13 (Type VI):** Targets RNA, not DNA—useful for transcriptome editing.

6.1.4. Advantages of CRISPR Technology

- **High Specificity and Efficiency**
- **Ease of Design and Use**
- **Multiplexing Capability:** Target multiple genes simultaneously.
- **Cost-Effectiveness:** Cheaper than previous genome-editing technologies like ZFNs and TALENs.

6.1.5. Limitations and Challenges

Despite its advantages, CRISPR has some limitations:

- **Off-Target Effects:** Unintended edits at similar sequences.
- **Delivery Systems:** Efficient and safe delivery into cells is still a challenge.
- **Ethical Considerations:** Particularly regarding germline editing and embryo modification.

6.2. Applications in Research and Therapy

CRISPR technologies have transformed biomedical research and therapeutic strategies by enabling precise manipulation of the genome. From functional genomics to gene therapy, CRISPR's versatility extends across model organisms, disease modeling, regenerative medicine, and even potential cures for genetic disorders.

6.2.1. Functional Genomics and Gene Knockout Studies

CRISPR has revolutionized **functional genomics** by allowing researchers to create targeted gene knockouts and study gene function.

- High-throughput CRISPR screens identify genes essential for cancer cell survival, drug resistance, or immune response.
- Enables systematic analysis of genetic pathways in various cell types and organisms.

6.2.2. Gene Therapy and Disease Correction

CRISPR is being developed to treat or even cure **inherited genetic diseases** by correcting mutations at the DNA level.

- Examples include **sickle cell anemia**, **beta-thalassemia**, and **Leber congenital amaurosis**.
- In vivo and ex vivo approaches are used depending on the tissue and target.

6.2.3. Stem Cell Engineering and Regenerative Medicine

CRISPR is widely used to genetically engineer **stem cells** for disease modeling and regenerative therapies.

- Induced pluripotent stem cells (iPSCs) edited with CRISPR can model patient-specific mutations.
- Enables creation of genetically corrected cells for transplantation.

6.2.4. Cancer Immunotherapy and CAR-T Cell Design

CRISPR facilitates **engineering immune cells** for cancer therapy.

- Used to design **Chimeric Antigen Receptor T cells (CAR-T)** with improved efficacy and reduced immune rejection.
- Enables knockout of immune checkpoint genes like PD-1 to enhance anti-tumor response.

6.2.5. Infectious Disease Research

CRISPR is applied in both diagnostics and treatment of infectious diseases:

- **CRISPR diagnostics (e.g., SHERLOCK, DETECTR)** enable rapid detection of viruses like SARS-CoV-2.
- Efforts are underway to use CRISPR to **inactivate latent viral genomes**, such as HIV provirus, in host cells.

6.2.6. Agricultural Biotechnology

In crop science, CRISPR is employed to:

- Enhance resistance to pests and diseases.
- Improve nutritional value and yield.
- Accelerate breeding without foreign DNA insertion (non-GMO).

Examples: Disease-resistant bananas, drought-tolerant rice, and non-browning mushrooms.

6.3. Ethical Considerations in Genome Editing

The advent of genome editing technologies, especially CRISPR-Cas systems, has ushered in transformative possibilities in science and medicine. However, the power to alter the genetic code of living organisms—especially humans—raises profound ethical, legal, and societal concerns. As genome editing moves from the laboratory into clinical and agricultural applications, a balanced dialogue involving scientists, policymakers, ethicists, and the public becomes increasingly essential.

6.3.1. Germline vs. Somatic Editing

A key ethical distinction is between **germline editing** (heritable changes in sperm, eggs, or embryos) and **somatic editing** (non-heritable changes in body cells).

- **Germline editing** affects future generations and is often viewed as more ethically contentious.
- **Somatic editing** is generally more accepted for treating diseases in living individuals.

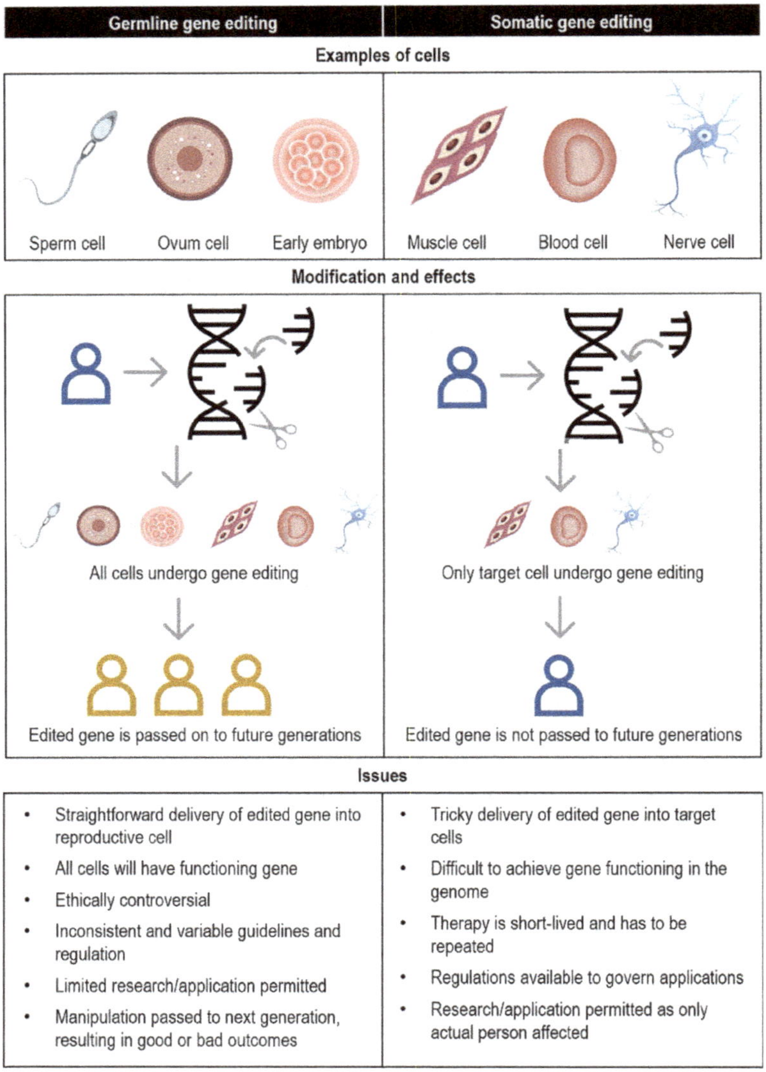

Figure 6.2: Illustration comparing germline editing (heritable) and somatic editing (non-heritable). Source: ResearchGate. Link: https://www.researchgate.net/figure/Comparison-between-germline-and-somatic-gene-editing-therapy-in-terms-of-the-types-of_fig1_357825083

6.3.2. Human Enhancement and "Designer Babies"

CRISPR raises the potential for **human enhancement**, such as altering genes to improve intelligence, physical abilities, or appearance—beyond medical necessity.

- Sparks debate over eugenics, equity, and the commodification of human life.
- Raises the question: where do we draw the line between therapy and enhancement?

6.3.3. Equity, Access, and Global Disparities

The benefits of genome editing may not be equally accessible across the globe or within countries:

- There is concern that advanced therapies will be available only to the wealthy.
- Disparities in **genomic research infrastructure** may exacerbate global health inequalities.

6.3.4. Informed Consent and Patient Autonomy

In clinical applications, **informed consent** is critical:

- Patients must understand the risks, benefits, and limitations of genome editing.
- Issues arise when editing is done on embryos or children who cannot consent.

6.3.5. Environmental and Ecological Risks

The use of genome editing in **ecosystems**—such as gene drives to eradicate mosquitoes—poses risks:

- Unintended consequences can disrupt ecological balance.
- Edited traits could spread uncontrollably in the wild.

6.3.6. Regulatory and Governance Challenges

Current regulatory frameworks are often outdated or inconsistent across countries:

- Some countries allow clinical gene editing; others ban it entirely.
- Global consensus is lacking on how to govern genome editing research and applications.

6.4. Off-target Effects and Precision Strategies

While CRISPR-Cas systems and other genome editing technologies offer unprecedented precision, they are not without flaws. A major concern is the possibility of **off-target effects**—unintended modifications at genomic sites other than the target sequence. These unintended changes may lead to unpredictable consequences, including gene disruption, mutations, or oncogenesis. To minimize these risks, researchers have developed a suite of **precision strategies** aimed at enhancing the accuracy and safety of genome editing.

6.4.1. Understanding Off-target Effects

Off-target effects occur when genome editing tools mistakenly bind and cleave at sequences similar, but not identical, to the intended target.

- These effects may be influenced by the **Cas protein type**, **gRNA sequence**, and **chromatin accessibility**.
- Unintended edits can cause **mutations**, disrupt regulatory regions, or activate oncogenes.

6.4.2. Detection Methods for Off-target Effects

Numerous experimental techniques are used to detect off-target edits:

- **GUIDE-seq** (Genome-wide Unbiased Identification of DSBs Enabled by Sequencing)
- **Digenome-seq**, **SITE-seq**, **CIRCLE-seq**, and **DISCOVER-seq**
- Whole-genome sequencing (WGS) for comprehensive off-target profiling

6.4.3. Improving Specificity with Engineered Cas Variants

To reduce off-target activity, researchers have developed **high-fidelity Cas variants**, including:

- **SpCas9-HF1**
- **eSpCas9** (enhanced specificity)
- **HypaCas9**
- **Sniper-Cas9**

These modified proteins maintain on-target activity while greatly reducing off-target cleavage.

6.4.4. Precision via Base Editing and Prime Editing

New genome editing tools aim to **rewrite DNA** without introducing double-strand breaks (DSBs):

- **Base Editors** (e.g., cytosine base editors, adenine base editors) convert one nucleotide to another without cutting the DNA.
- **Prime Editors** use reverse transcriptase and a specialized guide RNA to insert, delete, or replace specific DNA sequences.

These techniques offer more control and fewer off-target risks.

6.4.5. Bioinformatics Tools for gRNA Design

Computational tools help design guide RNAs (gRNAs) with minimal off-target potential:

- Examples include **CRISPOR**, **CHOPCHOP**, **Cas-OFFinder**, and **Benchling**.
- These tools analyze sequence context, mismatches, and potential off-target sites genome-wide.

Part III: Molecular Techniques in Modern Research

7. Next-Generation Sequencing and Omics Technologies

7.1. DNA, RNA, and Epigenome Sequencing

Next-generation sequencing (NGS) has revolutionized biological research by enabling massively parallel sequencing of millions of DNA or RNA molecules with high throughput and accuracy. This powerful tool supports a wide range of applications from genomic discovery to clinical diagnostics.

7.1.1. Principles of Next-Generation Sequencing

NGS technologies follow a general workflow:

1. **Sample Preparation** – Isolation of nucleic acids followed by fragmentation and adapter ligation.
2. **Amplification** – Clonal amplification (e.g., bridge PCR or emulsion PCR).
3. **Sequencing** – Parallel sequencing using platforms like Illumina, Ion Torrent, or PacBio.
4. **Data Analysis** – Alignment to reference genomes, variant calling, and interpretation.

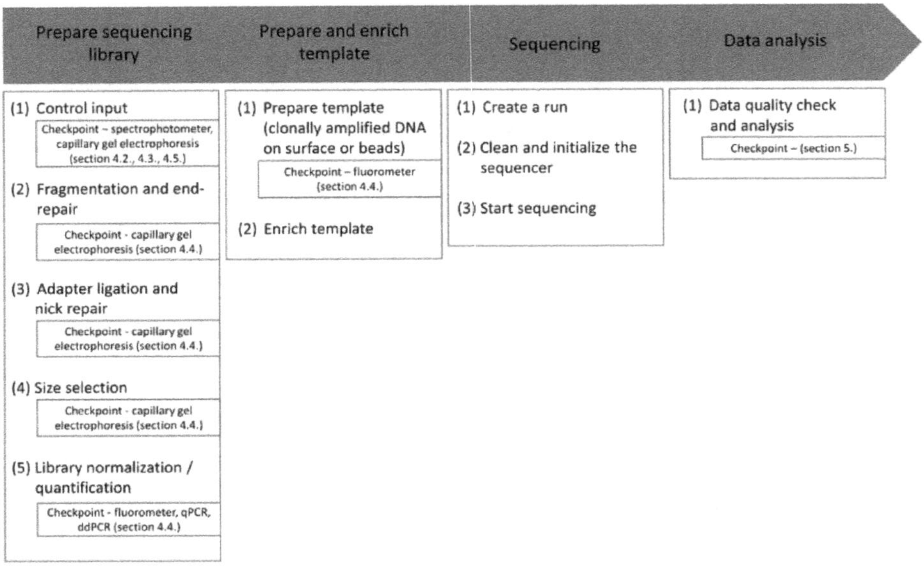

Figure 7.1: Overview of key steps in a typical NGS workflow including library prep, sequencing, and data analysis. Source: ResearchGate. Link: https://www.researchgate.net/figure/Overview-of-the-general-NGS-workflow-The-main-steps-library-and-template-preparation_fig1_304747820

7.1.2. DNA Sequencing Applications

NGS enables whole-genome sequencing (WGS), whole-exome sequencing (WES), and targeted sequencing:

- **WGS** reveals structural variants, mutations, and genome-wide alterations.
- **WES** focuses on coding regions, useful for detecting disease-causing mutations.
- **Targeted sequencing** allows deep sequencing of specific gene panels.

7.1.3. RNA Sequencing (RNA-seq)

RNA-seq profiles gene expression by sequencing transcriptomes:

- Quantifies gene expression levels.
- Detects novel transcripts, splicing events, and RNA editing.
- Enables single-cell RNA-seq (scRNA-seq) to study cellular heterogeneity.

7.1.4. Epigenome Sequencing

Epigenome sequencing investigates chemical modifications that regulate gene expression:

- **Bisulfite sequencing** maps DNA methylation at single-base resolution.
- **ChIP-seq** (Chromatin Immunoprecipitation sequencing) identifies histone modifications and transcription factor binding.
- **ATAC-seq** (Assay for Transposase-Accessible Chromatin) reveals chromatin accessibility.

7.1.5. Challenges and Future Directions

While NGS has transformed genomics, challenges remain:

- **Data volume and analysis complexity** demand powerful computational resources.
- **Sample quality**, library preparation, and sequencing biases can affect results.
- The future includes **real-time nanopore sequencing**, **integration of multi-omics data**, and **AI-based interpretation tools**.

7.2. Transcriptomics, Proteomics, and Metabolomics

High-throughput technologies have enabled system-level investigations into the molecular profiles of cells. **Transcriptomics**, **proteomics**, and

metabolomics are key pillars of **omics research**, allowing scientists to explore how genes are expressed, how proteins function, and how metabolites reflect cellular states.

7.2.1. Transcriptomics: Studying the Transcriptome

Transcriptomics refers to the comprehensive analysis of RNA transcripts in a cell or tissue at a specific time. It includes:

- **Bulk RNA-seq** for overall gene expression levels.
- **Single-cell RNA-seq (scRNA-seq)** to assess cell-to-cell variability.
- **Spatial transcriptomics** to localize gene expression in tissue context.

Applications:

- Identification of disease biomarkers.
- Functional annotation of genes.
- Discovery of new RNA isoforms and splicing events.

7.2.2. Proteomics: Profiling the Protein Landscape

Proteomics investigates the complete set of proteins (proteome) expressed in a cell, tissue, or organism. It reflects post-translational modifications and protein interactions that are not evident from transcript data.

Key methods:

- **Mass spectrometry (MS)-based proteomics**
- **2D gel electrophoresis**
- **Protein microarrays**

Applications:

- Mapping protein-protein interactions.
- Quantifying differential protein expression.
- Identifying post-translational modifications (e.g., phosphorylation, ubiquitination).

7.2.3. Metabolomics: Measuring Cellular Metabolites

Metabolomics focuses on the global profiling of small molecule metabolites such as sugars, amino acids, lipids, and nucleotides.

Techniques:

- Nuclear magnetic resonance (NMR) spectroscopy
- Gas chromatography–mass spectrometry (GC-MS)
- Liquid chromatography–mass spectrometry (LC-MS)

Applications:

- Characterization of metabolic pathways.
- Identification of metabolic biomarkers.
- Disease diagnosis and drug response monitoring.

7.2.4. Integrative Omics Approaches

Integrating transcriptomics, proteomics, and metabolomics enables a **systems biology** view of cellular processes:

- Reveals correlations and discrepancies between gene expression and protein/metabolite levels.
- Enhances precision in disease modeling and biomarker discovery.

- Enables **multi-omics network analysis** for pathway reconstruction and functional predictions.

7.3. Data Analysis and Bioinformatics Integration

Data analysis and bioinformatics are essential for making sense of vast amounts of genomic, transcriptomic, proteomic, and metabolomic data. This section explores the tools and strategies used to process and interpret complex biological datasets.

7.3.1. Bioinformatics Pipelines

- **Definition**: Bioinformatics pipelines are automated workflows that take raw data (e.g., from sequencing) through a series of computational steps, transforming it into meaningful biological information.
- **Components**: Pipelines typically include quality control, alignment, assembly, and variant calling.
- **Examples**: Pipelines like **Galaxy** and **Cufflinks** are commonly used for RNA-Seq data analysis.

7.3.2. Statistical Approaches in Omics Data

- **Data Normalization**: Before analysis, raw data must be normalized to remove biases and account for variability between samples.
- **Differential Expression Analysis**: In transcriptomics, tools like **DESeq2** and **edgeR** help in identifying genes that are differentially expressed under different conditions.
- **Clustering Techniques**: Methods such as **hierarchical clustering** and **k-means clustering** help identify patterns in omics data.

7.3.3. Integration of Multi-Omics Data

- **Overview**: Integrating multiple omics datasets (e.g., genomics, proteomics, and metabolomics) enables a more holistic understanding of biological systems.
- **Tools**: Software like **Cytoscape** and **Multi-Omics Factor Analysis (MOFA)** are widely used for integrative analysis.

7.3.4. Machine Learning in Bioinformatics

- **Application**: Machine learning algorithms (e.g., support vector machines, random forests) are applied to large datasets for predictive modeling.
- **Use Cases**: For example, **deep learning** is increasingly used in identifying biomarkers or predicting drug responses.

8. Molecular Cloning and Recombinant DNA Technology

8.1. Cloning Vectors and Host Systems

Molecular cloning is a fundamental technique in molecular biology used to replicate and manipulate DNA sequences in host organisms. This section focuses on the core components of cloning systems—vectors and hosts—that make gene manipulation and recombinant DNA technology possible.

8.1.1. What Are Cloning Vectors?

Cloning vectors are DNA molecules that carry foreign genetic material into a host cell, where it can be replicated and expressed.

Key Characteristics of Cloning Vectors:

- **Origin of replication (ori):** Ensures autonomous replication.
- **Selectable markers:** Commonly antibiotic resistance genes for identifying successful transformants.
- **Multiple cloning sites (MCS):** Regions with several restriction enzyme recognition sites to facilitate DNA insertion.

Common Types of Cloning Vectors:

- **Plasmids:** Circular DNA molecules used in bacterial cloning.
- **Bacteriophages (e.g., λ phage):** Used for cloning large DNA fragments.
- **Cosmids:** Hybrid vectors combining plasmid and phage features.
- **BACs (Bacterial Artificial Chromosomes):** Designed for cloning very large DNA fragments (up to 300 kb).
- **YACs (Yeast Artificial Chromosomes):** Used in yeast systems for cloning even larger fragments.

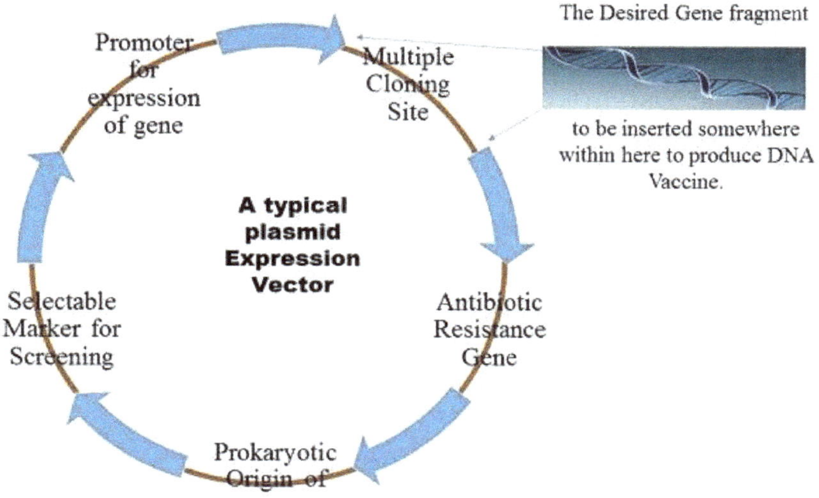

Figure 8.1: Structure of a Typical Plasmid Vector. Source: ResearchGate. Link: https://www.researchgate.net/figure/Schematic-diagram-of-a-typical-plasmid-expression-vector_fig1_320347914

8.1.2. Host Organisms in Cloning

The choice of host organism depends on the objective of cloning and the nature of the insert DNA.

Common Host Systems:

Prokaryotic Hosts (e.g., *Escherichia coli*):

- Fast growth rate
- High transformation efficiency
- Well-characterized genetics
- Limitation: Inability to perform complex post-translational modifications

Yeast (e.g., *Saccharomyces cerevisiae*):

- Eukaryotic system with simple genetic manipulation
- Capable of glycosylation and other eukaryotic modifications

Insect Cells (e.g., *Sf9* cells using baculovirus system):

- For expressing high yields of complex eukaryotic proteins

Mammalian Cells (e.g., HEK293, CHO cells):

- Suitable for therapeutic protein production
- Mimic human post-translational modifications

8.1.3. Selecting the Right Vector-Host System

The selection depends on:

- Size of DNA to be cloned
- Type of protein or gene product
- Desired post-translational modifications
- Speed, cost, and yield

Example Scenarios:

- Small gene cloning for expression in *E. coli*: Use plasmid vector (e.g., pUC19).
- Large genomic fragment for sequencing: Use BAC or YAC.
- Therapeutic protein requiring glycosylation: Use mammalian expression vector and CHO cells.

8.2. PCR, qPCR, and RT-PCR Techniques

Polymerase chain reaction (PCR) and its variants—quantitative PCR (qPCR) and reverse transcription PCR (RT-PCR)—are cornerstone molecular biology techniques for amplifying and analyzing DNA and RNA. These tools are indispensable for diagnostics, gene expression analysis, cloning, and more.

8.2.1. Polymerase Chain Reaction (PCR)

PCR Overview: PCR enables exponential amplification of specific DNA sequences using thermal cycling and DNA polymerase.

Key Components:

- Template DNA
- Primers (forward and reverse)
- DNA polymerase (e.g., Taq polymerase)
- dNTPs (deoxynucleotide triphosphates)
- Buffer and Mg^{2+}

PCR Cycle Steps:

1. **Denaturation (94–98°C):** DNA strands separate.
2. **Annealing (50–65°C):** Primers bind to target sequences.
3. **Extension (72°C):** Polymerase synthesizes new DNA strands.

8.2.2. Quantitative PCR (qPCR)

qPCR (Real-Time PCR): qPCR enables real-time monitoring of DNA amplification by using fluorescent dyes or probes.

Types of Detection:

- **SYBR Green:** Binds double-stranded DNA nonspecifically.
- **TaqMan Probes:** Sequence-specific, offering higher specificity.

Applications:

- Gene expression quantification
- Pathogen detection
- SNP analysis

8.2.3. Reverse Transcription PCR (RT-PCR)

RT-PCR Overview: RT-PCR converts RNA into complementary DNA (cDNA) using reverse transcriptase, followed by PCR amplification.

Steps:

1. RNA is extracted and purified.
2. Reverse transcription converts RNA to cDNA.
3. PCR amplifies the cDNA using specific primers.

Key Use Cases:

- Gene expression profiling
- Viral RNA detection (e.g., SARS-CoV-2)
- Alternative splicing analysis

8.2.4. Troubleshooting and Optimization

Common Issues:

- **Non-specific bands:** Primer-dimer formation or incorrect annealing temperature.
- **No amplification:** Poor template quality or missing reagents.
- **Low efficiency in qPCR:** Suboptimal probe/primer design.

Optimization Strategies:

- Gradient PCR to identify ideal annealing temperature
- Use of high-fidelity polymerases for cloning
- Ensuring RNA integrity for RT-PCR

8.3. Protein Expression and Purification Methods

Producing and isolating proteins is a fundamental part of molecular biology and biotechnology research. This section focuses on strategies for expressing recombinant proteins in various systems and purifying them for downstream applications such as structural studies, drug development, and enzyme assays.

8.3.1. Protein Expression Systems

1. Prokaryotic Systems (e.g., *E. coli*):

- **Advantages:** Fast growth, inexpensive, easy genetic manipulation.
- **Limitations:** Lack of post-translational modifications (PTMs), possible formation of inclusion bodies.

2. Eukaryotic Systems (e.g., yeast, insect, and mammalian cells):

- **Advantages:** Enable proper PTMs, folding, and secretion.
- **Limitations:** More complex and costly to maintain.

Expression Vectors:

- Include promoters (e.g., T7), antibiotic resistance genes, fusion tags (e.g., His-tag, GST-tag), and multiple cloning sites.

8.3.2. Induction and Optimization

Inducible Expression:

- **IPTG-Inducible Systems:** Common in *E. coli* using the lac operon.
- **Auto-induction Media:** Enable high-yield expression without manual induction.

Optimization Tips:

- Adjust temperature, induction timing, and IPTG concentration.
- Co-expression with chaperones to aid in proper protein folding.

8.3.3. Protein Purification Techniques

1. Affinity Chromatography:

- Uses tags like His_6 (binds to Ni-NTA resin), GST, or FLAG for selective binding and elution.
- High specificity and purity.

2. Ion Exchange Chromatography (IEC):

- Separates proteins based on charge.
- Can be used after affinity purification for further polishing.

3. Size Exclusion Chromatography (SEC):

- Separates proteins based on size; also helps remove aggregates or contaminants.

4. SDS-PAGE and Western Blotting:

- Used to verify protein expression and purity.

8.3.4. Solubility and Inclusion Body Management

Inclusion Bodies:

- Insoluble protein aggregates that often form in *E. coli*.
- Solubilized using denaturants (e.g., urea or guanidine), then refolded.

Improving Solubility:

- Lower expression temperature
- Use of fusion tags (e.g., MBP, NusA)
- Co-expression with folding chaperones

9. Advanced Microscopy and Imaging Techniques

9.1. Fluorescence and Confocal Microscopy

Microscopy has evolved from simple light imaging to powerful molecular-resolution techniques. Among the most widely used in cell and molecular biology are fluorescence and confocal microscopy—both of which allow visualization of specific cellular components and dynamic processes with high spatial resolution.

9.1.1. Principles of Fluorescence Microscopy

Fluorescence Microscopy relies on the emission of light by a fluorophore when excited by a specific wavelength. It allows visualization of specific molecules, structures, or processes in living or fixed cells.

Key Components:

- **Light Source:** High-intensity lamp or lasers.
- **Excitation/Emission Filters:** Select specific wavelengths.
- **Fluorophores:** Dyes or proteins like GFP, DAPI, FITC, Alexa Fluor.

Applications:

- Visualizing DNA (e.g., with DAPI), actin filaments, organelles, protein localization.
- Time-lapse imaging for dynamic studies.

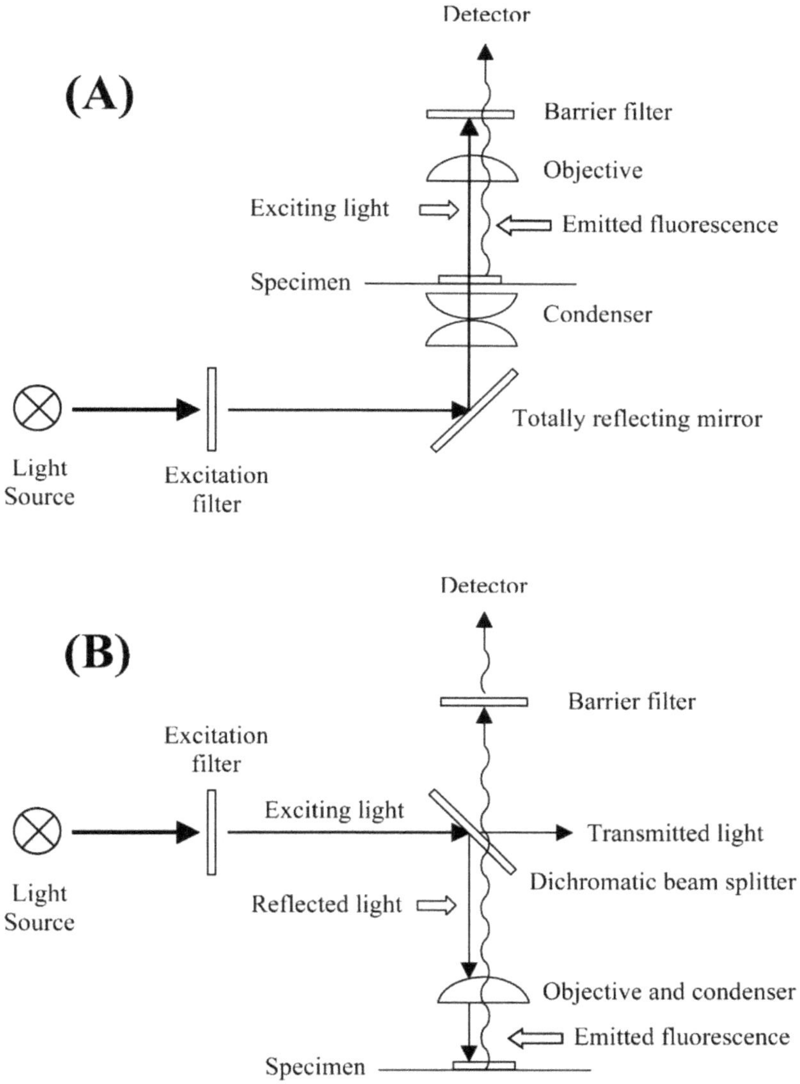

Figure 9.1: Schematic of a basic fluorescence microscope, showing excitation and emission pathways. Source: ResearchGate. Link: https://www.researchgate.net/figure/Light-paths-of-A-a-transmitted-light-fluorescence-microscope-and-B-an-incident-light_fig2_226662069

9.1.2. Confocal Laser Scanning Microscopy (CLSM)

Confocal Microscopy enhances resolution and contrast by eliminating out-of-focus light using a pinhole. It allows **optical sectioning**—producing sharp 3D reconstructions of specimens.

Advantages:

- Greater resolution and contrast than widefield fluorescence.
- Depth selectivity for thick specimens.
- Suitable for live-cell imaging and fixed tissue sections.

Mechanism:

- Uses point illumination via laser scanning.
- Emitted light passes through a pinhole to the detector.

9.1.3. Fluorescent Protein Tagging

Genetically Encoded Fluorescent Proteins like GFP, YFP, RFP allow visualization of gene expression and protein localization in real time.

Strategies:

- Fuse target gene with a fluorescent protein gene.
- Express in cells via plasmids or viral vectors.

Uses:

- Track protein movement and interactions.
- Study organelle dynamics and signaling events.

9.1.4. Multiplex Imaging and Co-localization

- **Multiplexing:** Use of multiple fluorophores to visualize different targets simultaneously.
- **Co-localization Analysis:** Determine if two molecules reside in the same region of a cell—critical for studying interactions.

9.2. Super-Resolution Imaging

Super-resolution microscopy techniques break the traditional diffraction limit of light (~200 nm), allowing researchers to visualize cellular structures at the nanometer scale. These methods have revolutionized our understanding of subcellular organization, protein complexes, and nanoscale molecular interactions.

9.2.1. The Diffraction Limit and Its Overcoming

According to Abbe's law, conventional optical microscopy is limited in resolution by the wavelength of light. Super-resolution imaging circumvents this using computational, optical, or stochastic techniques.

Techniques Used to Overcome the Limit:

- Structured illumination
- Temporal photon emission separation
- Computational image reconstruction

9.2.2. STED Microscopy (Stimulated Emission Depletion)

STED uses a depletion laser to quench fluorophores at the periphery of the excitation point, effectively shrinking the point spread function.

Key Features:

- High spatial resolution (~20–50 nm).
- Real-time imaging in live cells possible.
- Ideal for studying protein clusters and synaptic vesicles.

9.2.3. PALM and STORM (Single Molecule Localization)

PALM (Photo-Activated Localization Microscopy) and **STORM** (Stochastic Optical Reconstruction Microscopy) rely on temporally activating sparse subsets of fluorophores and localizing their center with high precision.

Advantages:

- Resolution ~10–30 nm.
- Excellent for studying protein complexes and membrane nanodomains.

Applications:

- Mapping nuclear pores, cytoskeletal filaments, receptor clustering.

9.2.4. SIM (Structured Illumination Microscopy)

SIM illuminates the sample with a grid-like pattern and shifts the pattern across the specimen. The resulting images are mathematically reconstructed for a higher-resolution composite.

Highlights:

- Doubles resolution (~100 nm).
- Suitable for live-cell imaging with low phototoxicity.

- Compatible with standard fluorophores.

9.2.5. Applications and Considerations

- **Applications:** Synaptic structure, nuclear architecture, chromatin organization, cytoskeletal dynamics.
- **Limitations:** Photobleaching, complexity of setup, computational load.
- **Best Use Cases:** When studying nanometer-scale structures where conventional fluorescence lacks resolution.

9.3. Live-Cell Imaging and Single-Cell Analysis

Live-cell imaging and single-cell analysis have transformed how we study dynamic biological processes in real time and at unprecedented resolution. These techniques allow researchers to observe individual cells as they interact, divide, differentiate, and respond to environmental cues—providing insight that population-level analyses cannot capture.

9.3.1. Principles of Live-Cell Imaging

Live-cell imaging involves tracking fluorescently labeled molecules or structures in living cells over time using specialized microscopes.

Core Features:

- Time-lapse microscopy with minimal phototoxicity.
- Environmental control (temperature, CO_2, humidity).
- Real-time visualization of dynamic processes.

9.3.2. Fluorescent Reporters and Biosensors

Live-cell imaging uses various fluorescent tools to visualize biological activity.

Types of Reporters:

- GFP-tagged proteins
- Calcium indicators (e.g., GCaMP)
- FRET biosensors for protein–protein interactions

Advantages:

- High specificity and spatiotemporal resolution.
- Suitable for multiplex imaging.

9.3.3. Applications in Cell Dynamics and Organelle Behavior

What It Reveals:

- Organelle trafficking (mitochondria, endosomes)
- Cytoskeleton remodeling
- Mitosis and cytokinesis
- Viral entry or drug response in real time

Use Cases:

- Cancer cell motility
- Neuron development
- Immune cell signaling

9.3.4. Single-Cell Analysis Techniques

Single-cell analysis uncovers heterogeneity within a cell population by analyzing genomic, transcriptomic, proteomic, or metabolomic data from individual cells.

Key Techniques:

- Single-cell RNA sequencing (scRNA-seq)
- Single-cell ATAC-seq
- Mass cytometry (CyTOF)

Applications:

- Identifying rare cell types
- Tracking developmental lineages
- Dissecting tumor microenvironments

9.3.5. Integration of Live Imaging with Single-Cell Omics

Combining live-cell imaging with downstream single-cell sequencing links phenotype to molecular profile.

Emerging Trends:

- Imaging-based cell sorting followed by omics analysis
- Use of barcoding techniques to trace live-imaged cells

Impact:

- Deep insight into cause-effect relationships in cell behavior.
- Enhanced understanding of cellular heterogeneity and transitions.

Part IV: Frontiers and Applications in Cell Science

10. Stem Cell Biology and Regenerative Medicine

10.1. Pluripotency and Differentiation

Understanding pluripotency—the capacity of a cell to develop into any cell type in the body—is fundamental to stem cell biology and regenerative medicine. The tightly regulated process by which stem cells differentiate into specialized cells underpins both normal development and therapeutic interventions.

10.1.1. Definition and Types of Stem Cells

Stem cells are categorized based on their potency and source:

- **Totipotent** – Can form all embryonic and extra-embryonic tissues (e.g., zygote).
- **Pluripotent** – Can differentiate into cells of all three germ layers (e.g., embryonic stem cells).
- **Multipotent** – Restricted to a particular lineage (e.g., hematopoietic stem cells).
- **Induced Pluripotent Stem Cells (iPSCs)** – Somatic cells reprogrammed back to a pluripotent state via transcription factors (e.g., Oct4, Sox2, Klf4, c-Myc).

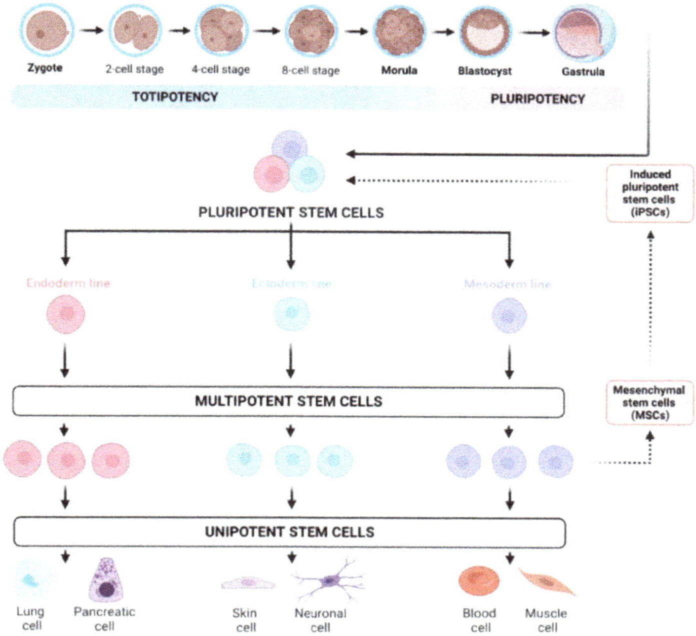

Figure 10.1: Overview of the differentiation potential of stem cells, from totipotent to unipotent states. Source: ResearchGate. Link: https://www.researchgate.net/figure/Levels-of-stem-cell-differentiation-potential-Totipotent-stem-cells-possess-the-highest_fig1_373097134

10.1.2. Molecular Basis of Pluripotency

Pluripotency is maintained through a network of transcription factors and epigenetic regulators:

- **Core Factors:** Oct4, Nanog, Sox2
- **Signaling Pathways:** Wnt, FGF, TGF-β
- **Epigenetic Features:** Open chromatin, bivalent histone marks, low DNA methylation

These elements work together to suppress differentiation while keeping the genome primed for lineage-specific gene expression.

10.1.3. Cellular Differentiation and Germ Layer Formation

Differentiation begins when pluripotent stem cells commit to one of the three germ layers:

- **Ectoderm:** Nervous system, skin
- **Mesoderm:** Muscle, bone, blood
- **Endoderm:** Gut, liver, pancreas

This commitment is driven by:

- External signals (growth factors, morphogens)
- Internal cues (lineage-specific transcription factors)
- Gradual epigenetic changes

10.1.4. Reprogramming Somatic Cells to iPSCs

The discovery of induced pluripotent stem cells (iPSCs) revolutionized stem cell biology by demonstrating that adult cells can be reprogrammed to pluripotency.

Key Concepts:

- **Reprogramming Factors:** Oct4, Sox2, Klf4, c-Myc (Yamanaka factors)
- **Techniques:** Viral vectors, non-integrating methods (e.g., mRNA, episomal vectors)
- **Applications:** Disease modeling, personalized medicine, drug screening

10.1.5. Ethical and Regulatory Considerations

The use of embryonic stem cells raises ethical concerns due to the destruction of embryos, whereas iPSCs offer a less controversial alternative. However, both still require strict regulatory oversight for:

- Genetic stability
- Tumorigenicity
- Informed consent and donor rights

10.2. Induced Pluripotent Stem Cells (iPSCs)

Induced pluripotent stem cells (iPSCs) have transformed regenerative medicine by enabling the reprogramming of adult somatic cells into a pluripotent state. iPSCs share key properties with embryonic stem cells, including self-renewal and the ability to differentiate into any cell type of the body. Their discovery opened new avenues for personalized medicine, disease modeling, and cell-based therapies without the ethical issues tied to embryonic sources.

10.2.1. Discovery and Development of iPSCs

iPSCs were first created in 2006 by Shinya Yamanaka through the introduction of four transcription factors—**Oct4, Sox2, Klf4, and c-Myc**— into mouse fibroblasts. This milestone demonstrated that mature cells can be epigenetically reprogrammed into a pluripotent state.

Key Milestones:

- 2006: Mouse iPSCs generated
- 2007: Human iPSCs developed
- Subsequent advances led to non-integrative reprogramming methods, improving clinical safety

10.2.2. Reprogramming Methods

The methods used to convert somatic cells into iPSCs can be divided into integrating and non-integrating techniques:

- **Integrating Methods:** Use of retroviruses or lentiviruses to deliver reprogramming genes; high efficiency but risk of genomic insertion
- **Non-integrating Methods:** Include Sendai virus, episomal plasmids, mRNA transfection, and small molecules; safer for clinical applications

10.2.3. Characterization and Validation of iPSCs

To confirm successful reprogramming, iPSCs are characterized through:

- **Morphological assessment**
- **Expression of pluripotency markers** (e.g., Nanog, TRA-1-60, SSEA-4)
- **Teratoma formation assay** (in vivo pluripotency test)
- **Embryoid body formation and in vitro differentiation**

10.2.4. Applications of iPSCs

iPSCs have broad applications in research and therapy:

- **Disease modeling:** Patient-specific iPSCs are used to study genetic disorders (e.g., ALS, Parkinson's)
- **Drug screening:** Platforms for testing toxicity and efficacy on human cells
- **Regenerative medicine:** Used to generate cardiomyocytes, neurons, or retinal cells for therapy
- **Personalized cell therapy:** Autologous transplantation reduces immune rejection

10.2.5. Limitations and Challenges

Despite their promise, iPSCs present several challenges:

- **Genomic instability** and potential for tumorigenesis
- **Incomplete reprogramming** and epigenetic memory
- **Low efficiency** and high cost in some reprogramming protocols
- **Standardization** and **regulatory hurdles** in clinical applications

10.3. Applications in Tissue Engineering

Tissue engineering integrates principles of biology, materials science, and engineering to restore, maintain, or enhance tissue and organ function. With the advent of stem cell technologies—particularly iPSCs—researchers can now create complex tissues in vitro for transplantation, drug screening, and disease modeling. The fusion of cell biology with biomaterials and bioprinting has accelerated advances in regenerative medicine and organ replacement.

10.3.1. Core Components of Tissue Engineering

Tissue engineering relies on three primary components:

- **Scaffolds:** Biocompatible structures that provide a 3D framework for cell attachment and tissue formation
- **Cells:** Typically stem cells or progenitor cells that can differentiate into target tissue types
- **Growth Factors:** Biochemical signals that guide cell differentiation, proliferation, and tissue development

10.3.2. Stem Cells in Tissue Engineering

iPSCs and other pluripotent stem cells offer a limitless source of patient-specific cells for creating personalized tissue grafts. Mesenchymal stem cells (MSCs) are also widely used due to their immunomodulatory properties and ease of isolation.

Applications:

- Skin regeneration using keratinocyte-derived sheets
- Cartilage and bone repair using MSC-laden hydrogels
- Myocardial patches derived from iPSC-cardiomyocytes

10.3.3. Biomaterials and Scaffold Design

Scaffolds must mimic the extracellular matrix (ECM) to support cell behavior and mechanical stability. Common materials include:

- **Natural polymers:** Collagen, fibrin, alginate
- **Synthetic polymers:** PLA, PLGA, PEG
- **Hybrid materials:** Combine synthetic strength with biological

functionality

Advanced scaffold designs incorporate **nanofibers**, **hydrogels**, and **bioactive coatings** to enhance cellular response and integration.

10.3.4. Bioprinting and Organ Fabrication

3D bioprinting enables layer-by-layer construction of tissues using bioinks containing living cells and biomaterials. This technology is pivotal in building vascularized tissues and complex organ structures.

Key Advancements:

- Bioprinted skin grafts for burn victims
- Vascularized cardiac patches
- Functional liver tissue models

10.3.5. Clinical Applications and Challenges

Tissue engineering has entered clinical practice in certain areas, such as skin substitutes and bladder reconstruction. However, challenges remain, including:

- **Vascularization** of thick tissues
- **Immune compatibility** in allogenic grafts
- **Long-term integration** and function
- **Regulatory approval** and manufacturing scale-up

11. Cancer Cell Biology and Molecular Oncology

11.1. Hallmarks of Cancer

Understanding cancer at the molecular level requires a framework that integrates the diverse biological capabilities acquired during tumor development. The concept of "hallmarks of cancer," first introduced by Hanahan and Weinberg (2000, updated 2011), outlines the fundamental traits that allow cancer cells to thrive, resist regulation, and metastasize. This framework has revolutionized molecular oncology by guiding research into cancer mechanisms and targeted therapies.

11.1.1. The Original Six Hallmarks

The foundational hallmarks include:

- **Sustaining proliferative signaling**: Cancer cells produce or respond excessively to growth signals.
- **Evading growth suppressors**: Tumors disable tumor suppressor pathways (e.g., p53, Rb).
- **Resisting cell death**: Mechanisms like apoptosis are circumvented.
- **Enabling replicative immortality**: Telomerase activation allows unlimited cell division.
- **Inducing angiogenesis**: Tumors stimulate new blood vessels for nutrient supply.
- **Activating invasion and metastasis**: Cells acquire motility and degrade ECM to spread.

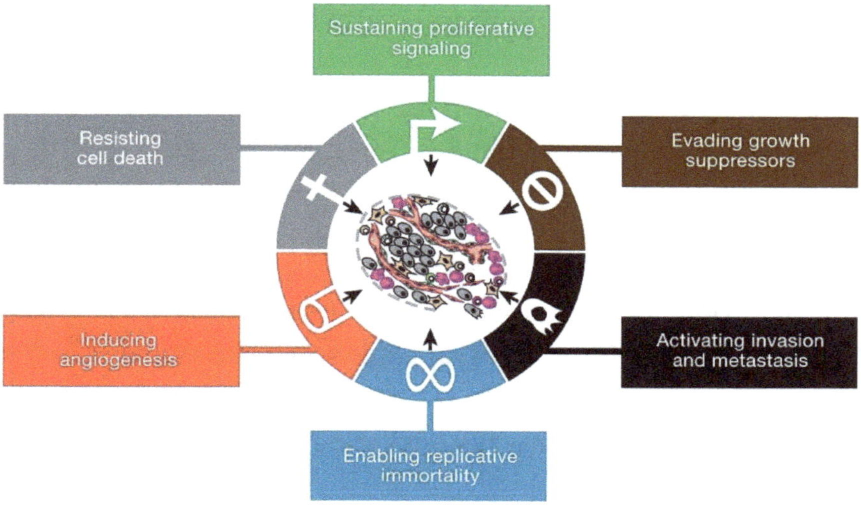

Figure 11.1: The six key capabilities that define the transformation of normal cells into malignant ones. Source: Cell. Link: https://www.cell.com/cell/fulltext/S0092-8674(11)00127-9?cc=y

11.1.2. Emerging Hallmarks and Enabling Characteristics

In their 2011 update, Hanahan and Weinberg introduced **emerging hallmarks** and **enabling characteristics**:

Emerging Hallmarks:

- **Reprogramming energy metabolism**: Cancer cells shift to aerobic glycolysis (Warburg effect).
- **Avoiding immune destruction**: Tumors evolve mechanisms to evade immune surveillance.

Enabling Characteristics:

- **Genome instability and mutation**: Increases mutation rates and drives evolution.
- **Tumor-promoting inflammation**: Chronic inflammation supports all

hallmarks.

11.1.3. Molecular Mechanisms Underlying the Hallmarks

Each hallmark is underpinned by genetic and epigenetic alterations, such as:

- **Oncogene activation** (e.g., RAS, MYC)
- **Tumor suppressor inactivation** (e.g., TP53, BRCA1)
- **Epigenetic silencing** of DNA repair or apoptotic genes
- **Noncoding RNAs** (e.g., miRNAs) influencing gene expression
- **Aberrant signaling pathways**, including PI3K-AKT, MAPK, and WNT

11.1.4. Clinical Relevance and Therapeutic Targeting

Understanding the hallmarks provides a roadmap for developing **targeted therapies** and **precision oncology**. Examples include:

- **EGFR inhibitors** for proliferative signaling (e.g., gefitinib)
- **Immune checkpoint inhibitors** for immune evasion (e.g., PD-1/PD-L1 blockers)
- **PARP inhibitors** for tumors with defective DNA repair (e.g., BRCA-mutated cancers)

11.2. Tumor Microenvironment and Metastasis

Cancer does not develop in isolation. The **tumor microenvironment (TME)**—composed of non-cancerous cells, extracellular matrix (ECM), and signaling molecules—plays a crucial role in promoting tumor progression and metastasis. Understanding the interactions within the TME and how they influence metastasis is vital for developing effective therapies.

11.2.1. Components of the Tumor Microenvironment

The TME includes a dynamic network of:

- **Cancer-associated fibroblasts (CAFs)** – Promote ECM remodeling and tumor growth
- **Immune cells** – Includes tumor-associated macrophages (TAMs), regulatory T cells (Tregs), and natural killer (NK) cells
- **Endothelial cells** – Drive angiogenesis to support tumor vascularization
- **Extracellular matrix (ECM)** – Provides structural support and regulates cell behavior
- **Signaling molecules** – Cytokines, chemokines, and growth factors (e.g., TGF-β, VEGF)

11.2.2. Tumor-Stroma Interactions and Inflammation

- Cancer cells interact with stromal cells through **paracrine signaling**, enhancing survival and invasion.
- **Chronic inflammation** in the TME promotes angiogenesis, DNA damage, and immune evasion.
- **TAMs and myeloid-derived suppressor cells (MDSCs)** release cytokines (e.g., IL-6, IL-10) that suppress cytotoxic immune responses.

11.2.3. Angiogenesis and Vascular Remodeling

- Tumors secrete **vascular endothelial growth factor (VEGF)** to stimulate new blood vessel formation.
- The **neo-vasculature** is often leaky and irregular, facilitating intravasation of tumor cells into the bloodstream.

- Anti-angiogenic therapies (e.g., bevacizumab) aim to disrupt this process.

11.2.4. Metastatic Cascade

The metastatic process includes:

- **Local invasion** – Tumor cells degrade ECM using enzymes like **matrix metalloproteinases (MMPs)**
- **Intravasation** – Entry of cancer cells into blood or lymphatic vessels
- **Circulatory survival** – Tumor cells evade immune attack and resist shear stress
- **Extravasation** – Exit from the bloodstream into distant tissues
- **Colonization** – Growth in a new microenvironment, forming secondary tumors

11.2.5. Pre-Metastatic Niche Formation

- Tumors can "prepare" distant sites via **exosomes** and **secreted factors** that condition local environments.
- Bone marrow-derived cells are recruited to support the formation of a **pre-metastatic niche**.

11.2.6. Clinical Implications and Therapeutic Targeting

Targeting the TME and metastasis includes:

- **Checkpoint inhibitors** to reactivate immune surveillance
- **Anti-angiogenic drugs** to inhibit vascular support
- **TGF-β and CXCR4 inhibitors** to disrupt tumor-stroma communication
- **MMP inhibitors** to block ECM degradation

These strategies aim to **impair the supportive role of the TME** and prevent metastatic dissemination.

11.3. Targeted Therapies and Drug Resistance Mechanisms

Advancements in molecular oncology have revolutionized cancer treatment through the development of **targeted therapies**. These treatments selectively attack molecular drivers of cancer with higher precision and fewer side effects compared to conventional chemotherapy. However, a major challenge is the development of **drug resistance**, which can render therapies ineffective over time.

11.3.1. Principles of Targeted Cancer Therapy

Targeted therapies are designed to interfere with specific molecular pathways that:

- Drive cancer cell proliferation and survival (e.g., EGFR, HER2, BCR-ABL)
- Promote angiogenesis (e.g., VEGF)
- Evade apoptosis (e.g., BCL-2)

Common types include:

- **Small-molecule inhibitors** (e.g., tyrosine kinase inhibitors)
- **Monoclonal antibodies** (e.g., trastuzumab, cetuximab)
- **Antibody-drug conjugates (ADCs)**

11.3.2. EGFR, HER2, and BCR-ABL Targeting

- **EGFR inhibitors**: Gefitinib, Erlotinib for NSCLC
- **HER2 inhibitors**: Trastuzumab for HER2+ breast cancer

- **BCR-ABL inhibitors**: Imatinib (Gleevec) for chronic myeloid leukemia

These therapies disrupt key oncogenic signals, leading to cancer cell death or growth arrest.

11.3.3. Mechanisms of Drug Resistance

Cancer cells develop resistance through:

- **Secondary mutations** (e.g., T790M in EGFR)
- **Gene amplification** (e.g., MET amplification)
- **Activation of alternative pathways** (e.g., bypass signaling)
- **Phenotypic plasticity** – transition to drug-tolerant persister cells
- **Efflux pump upregulation** – increased drug efflux by ABC transporters

11.3.4. Overcoming Resistance: Combination and Adaptive Therapies

Strategies include:

- **Combination therapy**: Targeting multiple pathways simultaneously to prevent compensatory activation.
- **Sequential therapy**: Altering treatment based on tumor evolution and resistance profile.
- **Adaptive therapy**: Modulating drug dosing to maintain a population of sensitive cells and delay resistance.
- **Second- and third-generation inhibitors**: Designed to overcome specific resistance mutations (e.g., Osimertinib for EGFR T790M).

11.3.5. Personalized Medicine and Biomarker-Driven Therapies

- **Next-generation sequencing (NGS)** allows detection of actionable mutations for precise drug selection.
- **Companion diagnostics** are tests used to identify patients likely to respond to specific treatments.
- Molecular profiling leads to **individualized therapy plans** with optimized efficacy and minimized toxicity.

12. Synthetic Biology and Cellular Engineering

12.1. Designing Genetic Circuits

Synthetic biology merges principles from biology and engineering to design and construct new biological parts, devices, and systems. A core component of this field is the **design of genetic circuits**—engineered networks of DNA sequences that control gene expression, often mimicking electronic circuits in their logic and behavior.

12.1.1. Introduction to Genetic Circuits

Genetic circuits are assemblies of **promoters**, **regulatory genes**, **repressors**, and **activators** that can produce predictable outputs in response to specific inputs. These circuits are designed to perform logical functions such as:

- Toggle switches (ON/OFF behavior)
- Oscillators (periodic gene expression)
- Logic gates (AND, OR, NOT)

They can be used for applications like biosensing, metabolic control, and programmable therapeutics.

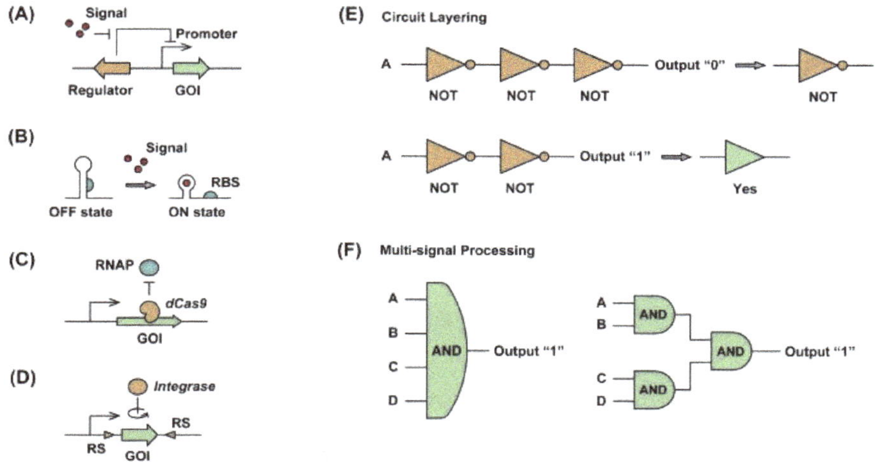

Figure 12.1: Examples of fundamental genetic circuit designs used in synthetic biology. Source: ScienceDirect. Link: https://www.sciencedirect.com/science/article/abs/pii/S0734975019300746

12.1.2. Components of a Genetic Circuit

- **Promoters**: Initiate transcription; can be constitutive or inducible.
- **Ribosome Binding Sites (RBS)**: Control translation initiation.
- **Coding Sequences (CDS)**: Encode functional proteins like enzymes or fluorescent reporters.
- **Regulatory Elements**: Transcriptional repressors, activators, feedback loops.
- **Terminators**: Signal end of transcription.

These components are often modular and standardized using systems like BioBrick or Golden Gate cloning.

12.1.3. Design and Modeling Approaches

Synthetic circuits are designed using computational tools to simulate gene expression dynamics and predict behavior. Popular modeling platforms include:

- **MATLAB SimBiology**
- **iBioSim**
- **CellDesigner**

Key parameters include promoter strength, degradation rates, and feedback strength.

12.1.4. Applications in Biosensing and Therapeutics

- **Biosensors**: Genetic circuits that detect environmental signals (e.g., toxins, pathogens) and generate measurable outputs.
- **Therapeutic circuits**: Engineered T-cells with circuits that recognize tumor antigens and control cytokine release.
- **Metabolic engineering**: Circuits regulate metabolic pathways for efficient production of pharmaceuticals and biofuels.

12.2. Biosensors and Biological Computing

Synthetic biology enables the creation of intelligent biological systems capable of sensing, processing, and responding to their environments. Two powerful applications of this approach are **biosensors**—cells or biomolecules that detect specific signals—and **biological computing**, which uses cells to perform logic operations similar to electronic circuits.

12.2.1. Biosensors: Overview and Design

Biosensors are engineered biological systems that convert the presence of a target molecule (e.g., toxin, pathogen, metabolite) into a measurable output, such as fluorescence or electrical signal.

Key components include:

- **Input detector** (e.g., receptor or aptamer)
- **Signal transducer** (e.g., regulatory protein or RNA switch)
- **Output reporter** (e.g., GFP, luciferase, color change)

Applications span environmental monitoring, disease diagnostics, food safety, and personalized medicine.

12.2.2. Cell-Based and Cell-Free Biosensors

- **Cell-based biosensors**: Use whole engineered cells (e.g., *E. coli*, yeast) to detect and respond to targets.
    - Useful for detecting antibiotics, heavy metals, or glucose levels.
- **Cell-free biosensors**: Rely on transcription-translation systems in vitro (e.g., paper-based sensors) without living cells.
    - Provide greater safety and portability.

12.2.3. Logic Gates in Biological Computing

Biological logic gates use molecules (DNA, RNA, or proteins) to mimic digital computation in living cells.

Types of biological logic gates:

- **AND gate**: Output occurs only when both inputs are present.
- **OR gate**: Output occurs if at least one input is present.
- **NOT gate**: Inverts the signal—no output when input is present.

Applications include:

- Programmable cancer therapy
- Smart diagnostics
- Molecular computing platforms

12.2.4. Real-World Applications

- **Environmental biosensing**: Detect pollutants like arsenic in groundwater using bacterial sensors.
- **Wearable diagnostics**: Paper-based biosensors embedded in fabrics to detect biomarkers in sweat.
- **Smart therapeutics**: Engineered probiotics that sense disease biomarkers and release therapeutic agents.

12.3. Future Applications in Medicine and Industry

Synthetic biology continues to unlock novel solutions across diverse sectors. Its integration into medicine and industry is reshaping how we approach diagnostics, therapeutics, biomanufacturing, and environmental sustainability. This section highlights the future potential of synthetic biology and cellular engineering in revolutionizing real-world applications.

12.3.1. Precision Medicine and Smart Therapeutics

Engineered cells and genetic circuits are being designed to perform **smart therapeutic functions**, such as detecting disease states and releasing treatments only when needed.

- **Programmable T-cells** for personalized immunotherapy (e.g., CAR-T therapy)
- **Gene circuits** that activate drugs based on environmental cues (e.g.,

inflammation, hypoxia)
- **Tumor-targeting probiotics** engineered to deliver cancer-killing agents directly into tumors

12.3.2. Industrial Biomanufacturing

Synthetic biology enables **bio-based production** of valuable materials, replacing chemical synthesis with more sustainable processes.

- **Biofuels**: Engineered algae and microbes producing ethanol, butanol, or biodiesel
- **Bioplastics**: Microbial production of biodegradable polymers (e.g., PHAs)
- **High-value compounds**: Synthesis of fragrances, flavors, vitamins, and pharmaceuticals

12.3.3. Environmental Sustainability and Bioremediation

Future applications include **green solutions** to environmental challenges using engineered biological systems.

- **Carbon capture**: Cyanobacteria engineered to enhance CO_2 fixation
- **Plastic degradation**: Bacteria designed to degrade PET and other plastics
- **Pollution sensors**: Cell-based biosensors to detect and neutralize heavy metals or oil spills

12.3.4. Agriculture and Food Innovation

Synthetic biology also promises transformative impacts on food systems.

- **Nitrogen-fixing cereals**: Engineering non-legumes to fix atmospheric nitrogen, reducing fertilizer dependency

- **Lab-grown meat**: Cellular agriculture producing meat without animals
- **Crop biosensors**: Plants that signal early signs of drought, pests, or nutrient deficiency

12.3.5. Challenges and Future Outlook

Despite its promise, challenges remain:

- Biosafety and ethical concerns about releasing engineered organisms
- Scalability and cost-efficiency in biomanufacturing
- Public perception and regulatory hurdles

Future focus areas include:

- Designing fail-safe genetic circuits
- Expanding the genetic code
- Building entirely synthetic minimal cells for novel applications

13. Cellular Aging and Senescence

13.1. Telomeres and Telomerase

Cellular aging is a fundamental biological process associated with the progressive decline in cellular function. At the heart of aging lies **cellular senescence**, a state in which cells cease to divide but remain metabolically active. This process plays dual roles—acting as a tumor suppressor mechanism and contributing to age-related tissue degeneration. One of the key molecular players in cellular aging is the **telomere–telomerase system**, which is explored in the following section.

Telomeres are protective caps located at the ends of chromosomes that prevent genomic instability. With each cell division, telomeres progressively shorten, ultimately triggering cellular senescence or apoptosis. Telomerase, an enzyme complex, plays a crucial role in extending telomeres and maintaining genomic integrity, particularly in stem cells, germ cells, and cancer cells.

13.1.1. Structure and Function of Telomeres

- Telomeres consist of repetitive DNA sequences (TTAGGG in vertebrates) and associated shelterin proteins that protect chromosome ends.
- These structures prevent chromosomal ends from being recognized as DNA breaks.
- Telomere shortening occurs due to the "end-replication problem" during DNA synthesis.

13.1.2. Telomerase Enzyme Complex

Telomerase is a ribonucleoprotein composed of:

- **TERT (telomerase reverse transcriptase)**: catalytic subunit
- **TERC (telomerase RNA component)**: RNA template used to extend telomeres

Active in germline cells, stem cells, and ~90% of cancer cells, but typically repressed in somatic cells.

13.1.3. Telomere Shortening and Replicative Senescence

- As cells divide, telomeres shorten until they reach a critical length, activating **DNA damage response pathways** (e.g., p53, p21).
- This leads to **replicative senescence**, a permanent growth arrest that limits the proliferative capacity of cells.
- This mechanism is essential for tumor suppression but contributes to tissue aging.

13.1.4. Telomerase Reactivation in Cancer

- In many cancers, **telomerase is reactivated**, allowing unlimited replication and cellular immortality.
- Telomerase inhibitors are being explored as **potential anticancer therapies**.
- Conversely, telomerase activation in stem cells may offer therapeutic potential for regenerative medicine.

13.2. Molecular Mechanisms of Aging

Aging is a complex, multifactorial process driven by a range of interconnected molecular pathways. These mechanisms contribute to the progressive decline of cellular and tissue function and are implicated in numerous age-associated diseases. Understanding these molecular underpinnings is essential for the development of targeted anti-aging therapies.

13.2.1. Genomic Instability

- **DNA damage accumulation** increases with age due to internal (e.g., reactive oxygen species) and external factors (e.g., radiation, chemicals).
- **Defective DNA repair pathways**, including base excision repair and non-homologous end joining, contribute to mutations and chromosomal aberrations.
- Genomic instability is a hallmark of both aging and cancer.

13.2.2. Epigenetic Alterations

- Aging is associated with **global hypomethylation** and **localized hypermethylation** of DNA, impacting gene expression.
- Histone modification patterns change, altering chromatin accessibility and transcriptional regulation.
- These epigenetic changes influence cell identity, inflammation, and disease susceptibility.

13.2.3. Loss of Proteostasis

- Proteostasis refers to the balance of protein folding, degradation, and synthesis.
- Aging impairs **proteasome function**, **autophagy**, and **chaperone-mediated folding**, leading to the accumulation of misfolded or damaged proteins.

- Protein aggregation is a hallmark of neurodegenerative diseases like Alzheimer's and Parkinson's.

13.2.4. Mitochondrial Dysfunction

- Mitochondria generate ATP and regulate apoptosis but also produce **reactive oxygen species (ROS)**.
- With age, **mitochondrial DNA (mtDNA) mutations** accumulate, impairing oxidative phosphorylation.
- Damaged mitochondria trigger inflammation and cell death, contributing to tissue degeneration.

13.2.5. Cellular Senescence and Inflammatory Signaling

- Senescent cells secrete **pro-inflammatory factors** known as the **senescence-associated secretory phenotype (SASP)**.
- Chronic low-grade inflammation or "**inflammaging**" exacerbates tissue damage and dysfunction.
- SASP components include IL-6, IL-8, and matrix metalloproteinases.

13.2.6. Deregulated Nutrient Sensing Pathways

Key signaling pathways involved in nutrient sensing include:

- Insulin/IGF-1 pathway
- AMPK (AMP-activated protein kinase)
- mTOR (mechanistic target of rapamycin)
- Sirtuins (SIRT1–7)

Caloric restriction and compounds like **rapamycin** or **resveratrol** target these pathways to extend lifespan in model organisms.

13.3. Anti-Aging Strategies and Therapeutics

The growing understanding of molecular aging mechanisms has accelerated the development of strategies aimed at delaying, preventing, or reversing age-related cellular decline. These interventions range from lifestyle-based approaches to cutting-edge molecular therapeutics.

13.3.1. Caloric Restriction and Fasting Mimicking Diets

- **Caloric restriction (CR)**—reducing calorie intake without malnutrition—has consistently shown to **extend lifespan** and **delay onset of age-related diseases** in multiple species.
- CR improves **mitochondrial function**, reduces **oxidative stress**, and regulates **insulin/IGF-1** and **mTOR** pathways.
- **Fasting mimicking diets (FMDs)** offer similar benefits by inducing metabolic reprogramming without continuous caloric reduction.

13.3.2. Senolytics and Senomorphics

- **Senolytics** are compounds that selectively **eliminate senescent cells**, reducing chronic inflammation and promoting tissue regeneration.
- Examples include **dasatinib**, **quercetin**, and **fisetin**.
- **Senomorphics** suppress the harmful secretory profile (SASP) of senescent cells without killing them.

13.3.3. NAD+ Boosters and Sirtuin Activators

- **NAD+ (Nicotinamide adenine dinucleotide)** levels decline with age, impairing energy metabolism and DNA repair.

- Compounds like **nicotinamide riboside (NR)** and **nicotinamide mononucleotide (NMN)** restore NAD+ levels.
- **Sirtuins**, especially **SIRT1**, are NAD+-dependent enzymes involved in lifespan extension and stress resistance.

13.3.4. mTOR Inhibitors and Rapalogs

- The **mTOR pathway** regulates growth, metabolism, and autophagy and is often hyperactive in aging cells.
- **Rapamycin** and its analogs (**rapalogs**) extend lifespan in animal models by inhibiting mTORC1, promoting autophagy, and reducing inflammation.
- Side effects and optimal dosing remain areas of ongoing investigation.

13.3.5. Telomerase Activation and Gene Therapy

- **Telomerase activation** can delay replicative senescence by extending telomeres, particularly in stem and immune cells.
- **Gene therapy approaches** aim to upregulate **TERT**, the catalytic subunit of telomerase.
- Concerns include **oncogenic potential** due to telomere-mediated immortalization.

13.3.6. Epigenetic Reprogramming and Yamanaka Factors

- Transient expression of **Yamanaka factors** (Oct4, Sox2, Klf4, c-Myc) can **partially reprogram aged cells**, reversing epigenetic aging without inducing pluripotency.

- This approach has restored **youthful function** in tissues such as retina, muscle, and pancreas in preclinical models.
- A promising frontier in **reversing biological age** without losing cell identity.

14. Cellular Responses to Environmental Stress

Cells constantly encounter environmental challenges such as oxidative stress, heat, toxins, and mechanical damage. Their ability to detect, respond to, and recover from these insults is vital for survival and function. This chapter focuses on the molecular response pathways—particularly oxidative stress management and heat shock protein expression.

14.1. Oxidative Stress and Heat Shock Proteins

14.1.1. Oxidative Stress and Reactive Oxygen Species (ROS)

- **Reactive oxygen species (ROS)** such as superoxide, hydrogen peroxide, and hydroxyl radicals are natural byproducts of cellular metabolism.
- When **ROS levels exceed antioxidant capacity**, oxidative stress occurs, leading to **DNA damage, protein oxidation**, and **lipid peroxidation**.
- Mitochondria are the primary site of ROS generation, especially during **electron transport chain (ETC)** dysfunction.
- Oxidative stress activates signaling pathways like **NF-κB, Nrf2**, and **p53**, influencing inflammation, cell cycle arrest, or apoptosis.

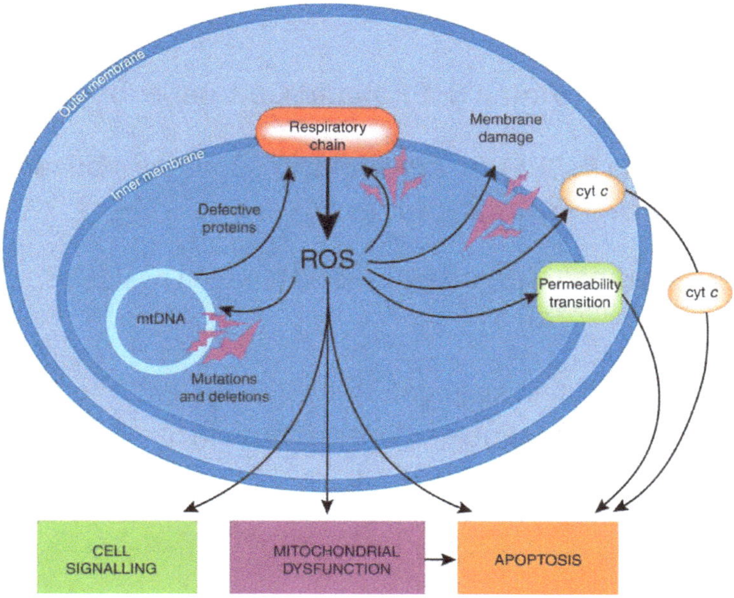

Figure 14.1: Overview of ROS production in mitochondria and its impact on cellular components. Source: ResearchGate. Link: https://www.researchgate.net/figure/Fig2-Overview-of-mitochondrial-ROS-production-ROS-production-within-mitochondria-can_fig3_51150692

14.1.2. Antioxidant Defense Mechanisms

- Cells deploy enzymatic antioxidants such as **superoxide dismutase (SOD)**, **catalase**, and **glutathione peroxidase (GPx)** to neutralize ROS.
- **Non-enzymatic antioxidants** include **vitamin C, vitamin E**, and **glutathione (GSH)**.
- **Nrf2** is a key transcription factor that regulates antioxidant gene expression. Under stress, Nrf2 translocates to the nucleus and activates genes with **antioxidant response elements (AREs)**.

14.1.3. Heat Shock Proteins (HSPs) and Cellular Protection

Heat shock proteins (HSPs) are molecular chaperones induced by heat, toxins, heavy metals, and oxidative stress.

Key HSPs include **HSP70**, **HSP90**, and **HSP27**, which assist in:

- Protein folding
- Prevention of protein aggregation
- Refolding or degradation of damaged proteins

Expression is controlled by **heat shock factor 1 (HSF1)**, which activates HSP transcription in response to stress.

14.1.4. Role of HSPs in Disease and Aging

- Dysregulation of HSP expression is linked to **neurodegenerative diseases** (e.g., Alzheimer's, Parkinson's), where misfolded proteins accumulate.

- In cancer, some tumors overexpress HSPs to evade apoptosis and enhance survival, making HSPs potential **therapeutic targets**.

- Aging cells often exhibit impaired stress response signaling, reduced HSP levels, and elevated oxidative damage.

14.2. Hypoxia, Nutrient Sensing, and Metabolic Adaptations

Cells must constantly monitor and adapt to changes in oxygen availability, nutrient levels, and energy demand. Hypoxia (low oxygen) and nutrient fluctuations trigger complex regulatory mechanisms that maintain homeostasis and support survival under stress. This section explores key molecular players in hypoxia signaling, nutrient sensing pathways, and metabolic adaptation strategies.

14.2.1. Hypoxia and the HIF Pathway

- **Hypoxia-inducible factors (HIFs)** are transcription factors central to the cellular response to low oxygen.
- Under normal oxygen (normoxia), **HIF-1α** is hydroxylated by prolyl hydroxylases and targeted for degradation by the **VHL E3 ubiquitin ligase**.
- In hypoxic conditions, hydroxylation is inhibited, stabilizing HIF-1α, which dimerizes with **HIF-1β**, translocates to the nucleus, and activates hypoxia-responsive genes.
- Targets include **VEGF (angiogenesis), EPO (erythropoiesis)**, and **glycolytic enzymes**.

14.2.2. Nutrient Sensing Pathways: mTOR and AMPK

- The **mTOR (mechanistic target of rapamycin)** pathway senses amino acids, glucose, and growth factors, promoting anabolic processes like protein synthesis and cell growth.
- In contrast, **AMP-activated protein kinase (AMPK)** is activated during low energy states (high AMP/ATP ratio), triggering catabolic processes to restore ATP levels.
- mTOR and AMPK exhibit **reciprocal regulation**, balancing growth and energy conservation.

14.2.3. Metabolic Reprogramming and Adaptation

- Hypoxia and nutrient stress often lead to **metabolic reprogramming**, where cells shift from oxidative phosphorylation to **aerobic**

glycolysis (Warburg effect).

- Cells upregulate **glucose transporters (GLUT1)** and **glycolytic enzymes** for ATP production under limited oxygen.
- **Autophagy** is also induced to recycle damaged organelles and provide internal nutrient sources during starvation.
- These adaptations are particularly important in **cancer cells, immune cells**, and **stem cells**, allowing survival and proliferation under harsh conditions.

14.2.4. Interplay with Inflammation and Disease

- Chronic activation of hypoxia and nutrient sensing pathways contributes to **inflammatory diseases, diabetes, obesity**, and **cancer**.
- Dysregulated mTOR activity is linked to **insulin resistance** and tumor growth.
- Hypoxia in the tumor microenvironment enhances **angiogenesis, immune evasion**, and **metastatic potential**.
- Therapeutic strategies aim to target these pathways using **HIF inhibitors, mTOR modulators**, and **AMPK activators**.

14.3. Immune and Inflammatory Responses

The immune system is essential for defending against pathogens and maintaining cellular homeostasis. In response to cellular stress, infection, or tissue damage, immune and inflammatory pathways are activated, involving a complex interplay of signaling molecules, immune cells, and gene regulation. This section explores the molecular mechanisms underlying immune activation and inflammation at the cellular level.

14.3.1. Innate Immunity and Pattern Recognition Receptors (PRRs)

- **Innate immunity** is the first line of defense and involves **pattern recognition receptors (PRRs)** that detect **pathogen-associated molecular patterns (PAMPs)** and **damage-associated molecular patterns (DAMPs)**.
- Key PRRs include **Toll-like receptors (TLRs)**, **NOD-like receptors (NLRs)**, and **RIG-I-like receptors (RLRs)**.
- Upon recognition, these receptors activate downstream signaling pathways like **NF-κB** and **interferon regulatory factors (IRFs)**, leading to the production of **cytokines**, **chemokines**, and **type I interferons**.

14.3.2. Pro-Inflammatory Cytokines and Signaling Pathways

- Inflammation is primarily driven by **pro-inflammatory cytokines**, including **TNF-α**, **IL-1β**, **IL-6**, and **IFN-γ**.
- These cytokines orchestrate recruitment of immune cells, promote fever, and upregulate adhesion molecules.
- **JAK-STAT** and **MAPK** signaling pathways are crucial for cytokine receptor activation and transcription of immune response genes.
- Dysregulation can lead to **chronic inflammation** and **autoimmune diseases**.

14.3.3. Inflammatory Cell Recruitment and Resolution

- Chemokines such as **CXCL8 (IL-8)** and **CCL2** attract **neutrophils, monocytes, and T cells** to sites of infection or injury.

- Endothelial cells express **selectins and integrins** to facilitate immune cell adhesion and extravasation.
- The resolution of inflammation involves **macrophage polarization (M1 to M2)**, **anti-inflammatory cytokines (e.g., IL-10)**, and **clearance of apoptotic cells**.

14.3.4. Cellular Crosstalk in Immune and Inflammatory Networks

- **Dendritic cells**, **macrophages**, **T cells**, and **epithelial cells** interact through cytokine networks and direct cell-cell contact.
- **Inflammasomes**, like **NLRP3**, activate caspase-1 and promote maturation of **IL-1β** and **IL-18**, amplifying the inflammatory response.
- **Crosstalk between immune and non-immune cells** in tissues like the gut, lungs, and skin regulates tissue-specific immunity and tolerance.

14.3.5. Implications in Disease and Therapeutic Targeting

- Chronic immune activation and unresolved inflammation are central to diseases such as **rheumatoid arthritis**, **atherosclerosis**, **inflammatory bowel disease (IBD)**, and **cancer**.
- Therapeutic strategies include **cytokine inhibitors (e.g., anti-TNF agents)**, **TLR antagonists**, and **inflammasome inhibitors**.
- Precision medicine approaches aim to target specific inflammatory pathways based on **molecular profiling**.

Appendix A: Common Laboratory Protocols

This appendix provides concise, step-by-step protocols for essential molecular and cell biology techniques. These procedures are widely used in research laboratories and serve as foundational tools for experimentation and data generation.

A.1. DNA Extraction (Phenol-Chloroform Method)

Purpose: Isolate high-quality genomic DNA from cells or tissues.

Materials:

- Lysis buffer (e.g., Tris-EDTA with SDS)
- Proteinase K
- Phenol:chloroform:isoamyl alcohol (25:24:1)
- Ethanol (100% and 70%)
- TE buffer

Protocol:

1. Lyse cells in lysis buffer; incubate with Proteinase K at 56°C for 1 hour.
2. Add equal volume of phenol:chloroform:isoamyl alcohol; vortex and centrifuge.
3. Transfer the aqueous phase to a new tube; repeat extraction if needed.
4. Precipitate DNA with cold 100% ethanol; wash with 70% ethanol.
5. Air-dry and resuspend DNA in TE buffer.

A.2. Agarose Gel Electrophoresis for DNA Analysis

Purpose: Separate DNA fragments by size.

Materials:

- Agarose powder
- TAE or TBE buffer
- Ethidium bromide or alternative stain
- DNA ladder
- Gel casting tray and comb

Protocol:

1. Prepare agarose gel (0.8–2% depending on fragment size).
2. Load DNA samples and ladder into wells.
3. Run gel at 80–120 V until adequate separation.
4. Visualize bands under UV or blue light.

A.3. Polymerase Chain Reaction (PCR)

Purpose: Amplify specific DNA sequences.

Materials:

- Template DNA
- Forward and reverse primers
- dNTPs
- Taq DNA polymerase
- PCR buffer
- Thermal cycler

Protocol:

1. Mix components: buffer, dNTPs, primers, polymerase, template.

2. Run PCR program:

2a. Initial denaturation: 95°C, 3 min

2b. 25–35 cycles of:

- Denaturation: 95°C, 30 sec
- Annealing: 50–65°C, 30 sec
- Extension: 72°C, 1 min/kb

2c. Final extension: 72°C, 5 min

3. Analyze product via agarose gel electrophoresis.

A.4. RNA Isolation (TRIzol Method)

Purpose: Extract total RNA from cells.

Materials:

- TRIzol reagent
- Chloroform
- Isopropanol
- Ethanol
- RNase-free water

Protocol:

1. Lyse cells in TRIzol; add chloroform and shake.
2. Centrifuge to separate phases.
3. Transfer aqueous phase to new tube; precipitate RNA with isopropanol.

4. Wash pellet with 75% ethanol; air-dry and resuspend in RNase-free water.

A.5. SDS-PAGE and Western Blotting

Purpose: Detect specific proteins in a sample.

Materials:

- SDS-PAGE gel
- Protein samples
- Transfer buffer
- PVDF/nitrocellulose membrane
- Primary and secondary antibodies
- Chemiluminescent substrate

Protocol:

1. Load proteins and run SDS-PAGE.
2. Transfer proteins to membrane.
3. Block membrane (e.g., with 5% milk).
4. Incubate with primary antibody, wash.
5. Incubate with HRP-conjugated secondary antibody.
6. Develop using substrate and image.

Appendix B: Glossary of Key Terms

This glossary provides concise definitions for key terms used throughout the book to assist in your understanding of advanced cell and molecular biology concepts.

A

- **Allele**: A variant form of a gene found at a specific locus on a chromosome.
- **Apoptosis**: Programmed cell death characterized by specific cellular morphological changes and biochemical events.
- **Autophagy**: A regulated cellular process in which cells degrade and recycle their own components via lysosomes.

B

- **Bioinformatics**: The application of computational tools to store, retrieve, analyze, and visualize biological data.
- **Bioreactor**: A vessel or device used to grow organisms under controlled conditions for biological or biochemical processes.

C

- **Chromatin**: The complex of DNA and proteins (mainly histones) in the nucleus that condenses to form chromosomes.
- **CRISPR**: A genome editing tool derived from bacterial immune systems, used to modify genes with precision.
- **Cyclins**: Regulatory proteins that control the progression of cells through the cell cycle by activating cyclin-dependent kinases (CDKs).

D

- **DNA Methylation**: An epigenetic mechanism involving the addition of methyl groups to DNA, affecting gene expression without changing the sequence.
- **Differentiation**: The process by which cells develop into distinct types with specialized functions.

E

- **Epigenetics**: The study of heritable changes in gene expression that do not involve changes in the DNA sequence.
- **Exon**: A segment of a DNA or RNA molecule containing information coding for a protein or peptide sequence.

G

- **Gene Expression**: The process by which genetic information is used to synthesize gene products (proteins or RNAs).
- **Genome**: The complete set of genetic material in an organism.

H

- **Histones**: Proteins around which DNA is wound in the nucleus; play a role in chromatin structure and gene regulation.
- **Hypoxia**: A condition in which tissue is deprived of an adequate oxygen supply.

M

- **Metabolomics**: The large-scale study of small molecules (metabolites) within cells, biofluids, tissues, or organisms.
- **Mitosis**: A type of cell division that results in two daughter cells with identical chromosomes to the parent cell.

O

- **Omics**: Collective technologies used to explore roles, relationships, and actions of the various types of molecules that make up the cells of an organism.

P

- **PCR (Polymerase Chain Reaction)**: A technique to amplify DNA sequences using cycles of denaturation, annealing, and extension.
- **Proteomics**: The large-scale study of proteins, particularly their structures and functions.
- **Pluripotent Stem Cells**: Cells that can develop into nearly all types of cells in the body.

R

- **RNA Interference (RNAi)**: A biological process in which RNA molecules inhibit gene expression by destroying specific mRNA molecules.
- **Ribosome**: A molecular machine in the cell that synthesizes proteins from messenger RNA templates.

S

- **Signaling Pathways**: A series of molecular interactions that transmit information from receptors to target molecules inside the cell.
- **Stem Cell**: An undifferentiated cell capable of giving rise to more cells of the same type and to differentiated cells.

T

- **Telomere**: The protective cap at the end of a chromosome that shortens with each cell division.
- **Transcriptomics**: The study of the complete set of RNA transcripts produced by the genome.

U

- **Ubiquitin**: A small protein that tags other proteins for degradation by the proteasome.

Z

- **Zygote**: A fertilized egg cell resulting from the fusion of a sperm and an ovum.

Appendix C: Online Resources and Databases

This appendix lists trusted online tools, databases, and platforms essential for students, educators, and researchers in advanced cell and molecular biology. These resources support data analysis, experimental planning, literature discovery, and continued learning.

1. National Center for Biotechnology Information (NCBI)

- **Website:** https://www.ncbi.nlm.nih.gov
- **Use:** Access to PubMed, GenBank, BLAST, and numerous genomic and protein databases.
- **Purpose:** Literature review, sequence alignment, gene information, and genetic variation.

2. UniProt (Universal Protein Resource)

- **Website:** https://www.uniprot.org
- **Use:** Protein sequence and functional annotation database.
- **Purpose:** Protein identification, structure, domains, and post-translational modifications.

3. Ensembl Genome Browser

- **Website:** https://www.ensembl.org
- **Use:** Genome data for vertebrates and model organisms.
- **Purpose:** Genomic exploration, gene expression, and comparative genomics.

4. KEGG: Kyoto Encyclopedia of Genes and Genomes

- **Website:** https://www.genome.jp/kegg/
- **Use:** Pathway maps for metabolism, signal transduction, and diseases.
- **Purpose:** Analyze and understand biological systems from molecular-level information.

5. GeneCards – The Human Gene Database

- **Website:** https://www.genecards.org
- **Use:** Detailed information on human genes.
- **Purpose:** Functional genomics, gene-disease associations, expression data.

6. EMBL-EBI (European Bioinformatics Institute)

- **Website:** https://www.ebi.ac.uk
- **Use:** Bioinformatics services including tools for protein domains, interactions, and alignments.
- **Purpose:** Integrative analysis of biological data.

7. Allen Brain Atlas

- **Website:** https://portal.brain-map.org
- **Use:** Gene expression maps in the brain.
- **Purpose:** Neurobiology studies and transcriptomics.

8. Human Protein Atlas

- **Website:** https://www.proteinatlas.org
- **Use:** Expression and localization of proteins in tissues and cells.
- **Purpose:** Protein biomarker discovery, immunohistochemistry.

9. UCSC Genome Browser

- **Website:** https://genome.ucsc.edu
- **Use:** Explore genomes with annotation tracks and comparative genomics tools.
- **Purpose:** Genomic visualization, gene mapping, evolutionary insights.

10. Reactome Pathway Database

- **Website:** https://reactome.org
- **Use:** Curated biological pathways and reactions.
- **Purpose:** Explore molecular interaction networks, signaling, and metabolism.

11. PDB (Protein Data Bank)

- **Website:** https://www.rcsb.org
- **Use:** 3D structures of biological macromolecules.
- **Purpose:** Structural biology, drug design, molecular docking.

12. Addgene Plasmid Repository

- **Website:** https://www.addgene.org
- **Use:** Non-profit plasmid repository and protocol sharing.
- **Purpose:** Molecular cloning, gene delivery, CRISPR tools.

13. Cell Signaling Technology (CST) Pathway Database

- **Website:** https://www.cellsignal.com/pathways
- **Use:** Signaling pathway diagrams and related reagents.
- **Purpose:** Experimental design and protein expression studies.

14. ResearchGate

- **Website:** https://www.researchgate.net
- **Use:** Academic network for accessing papers, discussions, and collaboration.
- **Purpose:** Connect with researchers, share data, ask questions.

15. PubMed Central (PMC)

- **Website:** https://www.ncbi.nlm.nih.gov/pmc/
- **Use:** Free access to full-text biomedical literature.
- **Purpose:** Literature review, reference mining, open-access journals.

Appendix D: Further Reading and Recommended Journals

This appendix provides a curated list of textbooks, review articles, and peer-reviewed journals that are valuable for deeper exploration of the topics covered in this book. These resources serve students, educators, and researchers seeking comprehensive insights or up-to-date advancements in the fields of cell and molecular biology.

A. Recommended Textbooks and Books

1. Molecular Biology of the Cell – Alberts, B. et al.

- A foundational reference in cell biology; known for its clarity and depth.
- ISBN: 978-0815344322

2. The Cell: A Molecular Approach – Cooper, G.M., & Hausman, R.E.

- Focuses on experimental approaches to understanding cell biology.
- ISBN: 978-1605352909

3. Lewin's Genes XII – Krebs, J.E., Goldstein, E.S., & Kilpatrick, S.T.

- Comprehensive book on gene structure and function.
- ISBN: 978-1284104493

4. Molecular Cell Biology – Lodish, H. et al.

- A rich resource combining molecular and cellular processes.
- ISBN: 978-1319114671

5. Essential Cell Biology – Alberts, B. et al.

- An approachable version of *Molecular Biology of the Cell* for beginners.
- ISBN: 978-0393884821

6. Epigenetics – Allis, C.D., Jenuwein, T., Reinberg, D.

- Covers foundational and emerging epigenetic mechanisms.
- ISBN: 978-1936113231

7. Principles of Gene Manipulation and Genomics – Primrose, S.B. & Twyman, R.M.

- Focuses on genetic engineering techniques and applications.
- ISBN: 978-1405135443

8. Stem Cells: From Biology to Therapy – Editors: Kursad Turksen

- In-depth exploration of stem cell biology and therapeutic potential.
- ISBN: 978-3030569246

9. Cancer Biology – Weinberg, R.A.

- A detailed examination of the molecular basis of cancer.
- ISBN: 978-0815342205

10. Synthetic Biology: A Primer – Tucker, J. & Zilinskas, R.

- Provides a broad overview of synthetic biology technologies.
- ISBN: 978-0262527736

B. Peer-Reviewed Journals

1. General Cell and Molecular Biology

- *Cell* – https://www.cell.com
- *Nature Cell Biology* – https://www.nature.com/ncb/
- *Journal of Cell Biology (JCB)* – https://jcb.rupress.org
- *Molecular Biology of the Cell (MBoC)* – https://www.molbiolcell.org

2. Genetics and Genomics

- *Nature Genetics* – https://www.nature.com/ng/
- *Genome Research* – https://genome.cshlp.org
- *BMC Genomics* – https://bmcgenomics.biomedcentral.com

3. Cancer and Oncology

- *Cancer Cell* – https://www.cell.com/cancer-cell
- *Nature Reviews Cancer* – https://www.nature.com/nrc/
- *Oncogene* – https://www.nature.com/onc/

4. Stem Cell Research

- *Stem Cell Reports* – https://www.cell.com/stem-cell-reports
- *Cell Stem Cell* – https://www.cell.com/cell-stem-cell
- *Stem Cells* – https://stemcellsjournals.onlinelibrary.wiley.com

5. Synthetic Biology and Systems Biology

- *ACS Synthetic Biology* – https://pubs.acs.org/journal/asbcd6
- *Nature Communications: Synthetic Biology Collection* – https://www.nature.com/collections/gjjcjcjbfc

6. Aging and Cellular Senescence

- *Aging Cell* – https://onlinelibrary.wiley.com/journal/14749726
- *The Journals of Gerontology: Series A* – https://academic.oup.com/biomedgerontology

References and Citation Sources

Foundational and Peer-Reviewed References

Alberts, B., Johnson, A., Lewis, J., Raff, M., Roberts, K., & Walter, P. (2014). *Molecular biology of the cell* (6th ed.). Garland Science.

Lodish, H., Berk, A., Kaiser, C. A., Krieger, M., Bretscher, A., Ploegh, H., ... & Darnell, J. (2021). *Molecular cell biology* (9th ed.). W.H. Freeman.

Cooper, G. M., & Hausman, R. E. (2019). *The cell: A molecular approach* (8th ed.). Sinauer Associates.

Lewin, B. (2020). *Genes XII*. Jones & Bartlett Learning.

Pierce, B. A. (2020). *Genetics: A conceptual approach* (7th ed.). W.H. Freeman.

Berg, J. M., Tymoczko, J. L., Gatto, G. J., & Stryer, L. (2019). *Biochemistry* (9th ed.). W.H. Freeman.

Watson, J. D., Baker, T. A., Bell, S. P., Gann, A., Levine, M., & Losick, R. (2013). *Molecular biology of the gene* (7th ed.). Pearson.

Allis, C. D., Caparros, M. L., Jenuwein, T., & Reinberg, D. (2015). *Epigenetics* (2nd ed.). Cold Spring Harbor Laboratory Press.

Doudna, J. A., & Sternberg, S. H. (2017). *A crack in creation: Gene editing and the unthinkable power to control evolution*. Houghton Mifflin Harcourt.

Lodish, H. (2021). *Introduction to recombinant DNA technology and applications*. In *Molecular cell biology* (9th ed.). W.H. Freeman.

Weinberg, R. A. (2013). *The biology of cancer* (2nd ed.). Garland Science.

Lander, E. S. (2011). Initial impact of the sequencing of the human genome.

Nature, 470(7333), 187–197. https://doi.org/10.1038/nature09792

He, L., & Hannon, G. J. (2004). MicroRNAs: Small RNAs with a big role in gene regulation. *Nature Reviews Genetics*, 5(7), 522–531.

van der Oost, J., Westra, E. R., Jackson, R. N., & Wiedenheft, B. (2014). Unravelling the structural and mechanistic basis of CRISPR–Cas systems. *Nature Reviews Microbiology*, 12(7), 479–492.

Bassett, A. R., & Liu, J. L. (2014). CRISPR/Cas9 mediated genome engineering in Drosophila. *Methods*, 69(2), 128–136.

Wikipedia and Wikimedia Sources

Wikipedia contributors. (2024). *CRISPR*. Wikipedia. https://en.wikipedia.org/wiki/CRISPR

Wikipedia contributors. (2024). *Cell signaling*. Wikipedia. https://en.wikipedia.org/wiki/Cell_signaling

Wikipedia contributors. (2024). *Heat shock protein*. Wikipedia. https://en.wikipedia.org/wiki/Heat_shock_protein

Wikipedia contributors. (2024). *Induced pluripotent stem cell*. Wikipedia. https://en.wikipedia.org/wiki/Induced_pluripotent_stem_cell

Wikimedia Commons contributors. (2024). *Illustrations of molecular biology*. Wikimedia Commons. https://commons.wikimedia.org/wiki/Category:Molecular_biology

ResearchGate and Open Access Sources

Zhao, Y., et al. (2020). Telomere shortening and aging: Mechanisms and therapeutic targets. *Aging and Disease*. https://www.researchgate.net/publication/339147230

Khan, I., et al. (2021). Role of autophagy and apoptosis in cancer therapy. *ResearchGate*. https://www.researchgate.net/publication/350300002

Liu, Y., & Cao, X. (2016). Immunosuppressive cells in tumor immune escape and metastasis. *Journal of Molecular Cell Biology*. https://www.researchgate.net/publication/305977416

Qadir, M. I., & Malik, S. A. (2010). Mechanism of gene silencing by RNA interference. *Biotechnology and Molecular Biology Reviews*, 5(9), 171–177. https://www.researchgate.net/publication/228668028

Mustafa, S. B. (2020). Synthetic biology: Future applications in industry. *ResearchGate*. https://www.researchgate.net/publication/344050881

Topic-Specific Journal Articles

Macosko, E. Z., et al. (2015). Highly parallel genome-wide expression profiling of individual cells using nanoliter droplets. *Cell*, 161(5), 1202–1214.

Choudhuri, S. (2014). *Bioinformatics for beginners: Genes, genomes, molecular evolution, databases and analytical tools*. Academic Press.

Collins, F. S., & Varmus, H. (2015). A new initiative on precision medicine. *New England Journal of Medicine*, 372(9), 793–795.

Chari, R., Mali, P., Moosburner, M., & Church, G. M. (2015). Unraveling CRISPR-Cas9 genome engineering specificity using in vitro assays. *Nature Methods*, 12(7), 607–614.

Ghosh, S., & Tergaonkar, V. (2007). NF-κB and cellular responses to stress. *Science's STKE*, 2007(432), re8. https://doi.org/10.1126/stke.2007.432.re8

Advanced Cell and Molecular Biology

A Deep Dive into Cutting-Edge Research, Molecular Techniques, and Emerging Frontiers in Cell Science

Published in the United States

ISBN: 9798316167340

Disclaimer

www.ingramcontent.com/pod-product-compliance
Lightning Source LLC
LaVergne TN
LVHW060037210926
848481LV00013B/3886
* 9 7 9 8 3 4 9 2 5 1 0 0 9 *